DESIGNING FOR QUALITY

DESIGNING FOR QUALITY

AN INTRODUCTION TO THE BEST OF TAGUCHI AND WESTERN METHODS OF STATISTICAL EXPERIMENTAL DESIGN

ROBERT H. LOCHNER
Consultant in Statistics and Quality Improvement

JOSEPH E. MATAR
Professor of Mechanical and Industrial Engineering
Marquette University

Quality Resources
A Division of The Kraus Organization Limited
White Plains, New York

ASQC Quality Press
American Society for Quality Control
Milwaukee, Wisconsin

Printed in the United States of America

94 93 92 91 90   10 9 8 7 6 5 4 3 2 1

Quality Resources
A Division of The Kraus Organization Limited
One Water Street, White Plains, New York 10601

Library of Congress Cataloging-in-Publication Data

Lochner, Robert H.
    Designing for quality : an introduction to the best of Taguchi and
  western methods of statistical experimental design / Robert H.
  Lochner and Joseph E. Matar.
       p.   cm.
    Includes bibliographical references (p.   ).
    ISBN 0-527-91633-1
    1. Taguchi methods (Quality Control)   2. Experimental design.
  I. Matar, Joseph E.   II. Title.
  TS156.L62   1990
  658.5′62—dc20   90-33724
                                                  CIP

To Sarajane and Jean—

for making the good times better,
for making the hard times easier,
for making our lives so much richer.

R.H.L. and J.E.M.

Contents

Preface

When Dr. Genichi Taguchi first brought his ideas to America in 1980 he was already well known in Japan for his contributions to quality engineering. His arrival in the U.S. went virtually unnoticed, but by 1984 his ideas had generated so much interest that Ford Motor Company sponsored the first Supplier Symposium on Taguchi Methods. Some people consider Dr. Taguchi the most important contributor to quality engineering concepts and methods. Others say that his experimental designs had already been developed by others and that he added nothing new of value. Problems of translation from Japanese to English added to the confusion. Throughout much of the 1980s the debate continued, and has only recently subsided. What seems clear now is that Dr. Taguchi has added greatly to our understanding of the importance of properly designed engineering experiments in quality improvement and has provided a structural approach to designing quality into products and processes. His experimental designs are, for the most part, not new. Nor was he the first to use a quadratic loss function or signal-to-noise ratios. But his use of experimental designs to reduce variability and make designs more robust had a major influence on perceptions of the role of experimental design in quality improvement.

This book does not dwell on the controversies regarding Taguchi Methods. Nor is this book exclusively about Taguchi Methods. For example, the important work of Dr. George Box and his associates is the basis for much of the material on orthogonal designs. The purpose of the book is to show engineers with little or no previous exposure to experimental design how to use statistically designed experiments to improve products and processes. A prerequisite for this book is an appreciation for the concept of random variation, as is gained through a typical statistics course or by using Statistical Process Control (SPC) in a manufacturing environment.

The authors are grateful to Don Clausing, Thomas C. Hsiang, and J. Richard Zelonka for their helpful comments on the manuscript. We also thank Randall F. Culp, who shared his ideas with us, leading to a simplified approach to calculation of estimated effects. Bill Golomski was a valuable teacher for the first author, patiently showing him ways in which quality improvement and statistics are intertwined. Lisa Budde did an outstanding job of converting our crude drawings into finished figures. The staff of Quality Resources provided support and encouragement to us at those critical times when we were despairing of ever finishing the book. Finally, we are most grateful to our families, particularly Sarajane and Jean, for their patience with us during those countless long evenings and weekends we spent working on this manuscript rather than painting the house, cleaning the garage, or sailing.

R. H. Lochner
J. E. Matar
Milwaukee, Wisconsin

Introduction

1

1.1 Modern Quality Control

As recently as twenty years ago most Quality Control (QC) departments were inspection operations. The head of the department often had the title Chief Inspector. Today quality control, or quality assurance, is much more than inspection. Some companies have even delegated most of the inspection responsibilities to production workers, leaving the quality professionals free to concentrate on quality improvement activities. At other companies quality control people still spend most of their time looking for defective products. This is a poor use of human resources. QC people should be helping to prevent problems rather than just reacting to them.

What does a modern quality system look like? First of all, it is *customer focused*. Products are designed and produced and services provided to meet customer requirements. No more creating markets for new gadgets. There is also a recognition of internal (to the company) as well as external customers. Everyone in the company identifies who their customers and suppliers are and what they require. Managers are customers of their secretaries, but they are also suppliers to their secretaries. If a report needs to be prepared, it is the responsibility of the manager (supplier) to get the material to the secretary (customer) so that the report can be properly prepared and sent to the person who needs the report (another customer). The practice of each department optimizing its own operations, without regard to the needs of other departments in the company, is no longer acceptable.

Another characteristic of a modern quality system is that the quality improvement process is *led by top management*. If the responsibility for quality is delegated to the QC department, then everyone in the company understands that quality is not a key management concern. There must be active, visible involvement by top management in the quality effort. Former Secretary of Labor William E. Brock said that "Most workers want to do a good job, want to turn out a quality product, want to provide a quality service, and want to be proud of what they do. But the tone must be set with management. If quality is not on management's agenda, it isn't going to be on the workers'.

If workers perceive that management's only interest is in what they describe as the corporate bottom line, then their interest in turn will only be in their bottom line—wages and benefits" (National Quality Forum III, as reported in *Quality Progress,* February, 1988, page 38).

It is important that *everyone understand their specific responsibilities for quality.* Although it is true that "Quality is everyone's job," it is also true that everyone has a different responsibility, depending on his or her position within the company. Product Development is responsible for designing new products which meet customer requirements and which can be consistently and economically produced by Manufacturing (Manufacturing is a customer of Product Development). Purchasing must not buy materials on price and delivery alone. Top management must show by word and action that quality is of utmost importance to the company's survival. Dr. W. Dekker, President and Chairman of the Board of Management of Philips Corporation, one of the world's largest manufacturing companies, said that "the quality of products and services is of the utmost importance for the continuity of our company. By adopting a quality policy aimed at complete control of every activity, maximum quality, productivity, flexibility and a reduction in cost . . . will be achieved. Every employee must be imbued with an attitude directed towards a continuous striving for improvements" (quoted in Filley, 1985, pages 98–99).

A fourth characteristic of a modern quality system is that it is *defect prevention oriented* rather than being focused on defect detection. Quality through inspection is not enough, and is too expensive. Although it may be necessary to do some sort of inspection or audit of finished products, a company's quality efforts should be focused on preventing defects from occurring—on doing the right thing right the first time. A modern quality system works on controlling processes rather than focusing on finished products. If the process is right, the finished product will be right. David T. Kearns, Chairman and CEO of Xerox Corporation, said that "quality control . . . has nothing to do with inspecting a product at the end of the assembly line. It is the involvement of cost and quality in every function, and the awareness that one drives the other. High quality drives your costs down" (National Quality Forum III, as reported in *Quality Progress*, February, 1988, page 29).

Finally, *quality must be a way of life.* Quality issues are discussed at every management meeting; there is no consideration given to shipping an unacceptable product in order to meet production schedules. Old technology is replaced by new only when the new technology is proven to be dependable. All employees are given training in modern quality concepts and methods. Everyone in the company is allowed to participate in the quality improvement effort. John E. Young, President and CEO of Hewlett-Packard, said that "quality does not come from simply having a large quality assurance staff to weed out defective products or services before they get to the customer; it is achieved by making the improvement process a permanent part of a company's culture" (*Fortune*, October 14, 1985).

A modern quality system can be divided into two parts:

- Quality of design
- Quality of conformance

Quality of design refers to those activities which assure that new or modified products and services are designed to meet customer needs and expectations, and are economically achievable. Quality of design is primarily the responsibility of Research and Development (R&D), Process Engineering, Market Research, and related groups.

Quality of conformance refers to manufacturing products or providing services which meet previously determined and clearly defined specifications. Production, Scheduling, Purchasing, and Shipping have primary responsibility for quality of conformance. Management has the responsibility to see that people have the training, tools, and other resources to do their jobs, and have the opportunity to become involved in the quality improvement process.

1.2 Quality in Engineering Design

"Quality cannot be inspected out of a product, it must be built in." We hear this statement so often that we might start to think that everyone is busy doing just that—designing and building quality into products, processes, and services. But the reality is that few people know how to "build quality." What passes for "building" in many companies is problem solving. Problem solving is, of course, an important activity. But wouldn't it be better if we could design products and processes so that the problems never occurred in the first place? James E. Olson, late President and CEO of AT&T, said, "We believe that quality is not something you add to a product, rather something you put into a product or service right from design on through customer feedback. When people say that quality is going to cost money—I don't believe that. When you do it right the first time, you'll end up with a satisfied customer and in fact reduced costs" (*Quality*, October, 1985, page 20). A pipe dream? Not really. There are companies where problems are prevented at the design phase, where product quality is addressed before the specifications are issued. At these companies some or all of the following six methods are used by design and development engineers:

- Concurrent engineering
- Quality function deployment (QFD)
- Reliability analysis
- Failure mode and effects analysis (FMEA)
- Taguchi Methods®
- Statistically designed experiments

Concurrent engineering refers to the integrated development of products, their manufacturing processes, and support systems. Rather than have Research and Development design a new product, pass the new design on to manufacturing engineers to develop manufacturing procedures, and then turn the new process over to manufacturing, manufacturing and support personnel work together under concurrent engineering design. The result is reduced development and manufacturing costs, and a product that meets customer expectations. Concurrent engineering is more an engineering philosophy than a method. It is similar to systems engineering and simultaneous engineering (see Winner et al., 1988).

Quality function deployment was developed at the Kobe Shipyard in Japan in the early 1970s. It has been used successfully by many Japanese firms, most notably Toyota. QFD is a highly structured technique for assuring that the "voice of the customer" is not lost in the noise of product development. Sullivan (1986) describes the following four basic components:

- Planning matrix—The rows of this matrix represent customer requirements in customer terms, and the columns list final product control characteristics. Marks made within the matrix represent relationships between customer requirements and product control characteristics.
- Deployment matrix—Customer requirements and product control characteristics are listed in greater detail in this matrix, so that critical component characteristics can be identified.
- Process plan and quality control charts—The process plan chart shows the relationship between manufacturing processes and critical component characteristics. It aids in identifying where monitoring points for the process are most needed. The quality control chart includes a schematic of the process flow and describes the necessary control points by location, type, and frequency of use.
- Operation instructions—This document describes operations which plant personnel must perform to assure that critical product requirements are met.

QFD is time-consuming, but the added work "up front" can mean significant savings later and increased customer satisfaction with the product.

Reliability analysis addresses the quality of a product over time. Will the product operate properly during the required or anticipated time of use? Statistical analysis of failure data, life testing, probability models of survival, and FMEA are all important tools in reliability engineering.

Failure mode and effects analysis is a tool used by design engineers and design review committees to protect against potential problems in processes and products. FMEA is a systematic approach to identifying potential product or process failures (failure modes), their effects in terms of functionality and safety, and what steps need to be taken to protect against these failures. FMEA should be done during product development, and again when unanticipated failures occur or there are design changes. Timely use of FMEA can avoid expensive modifications by revealing potential design deficiencies or hazards prior to actual production. Usually an FMEA is carried out from the bottom up, starting at the component level and working up through subsystems to the complete system. FMEA should examine performance under extreme environmental conditions and should try to anticipate possible misuse or misapplication by customers. FMEA was originally used in the aerospace industry to prevent potentially catastrophic equipment failures. The technique is now applied to production processes and service operations as well. Food and drug industries use FMEA under the name Hazard Analysis and Critical Control Points (HACCP) as a way of preventing contamination of products during production and handling.

Taguchi Methods refer to techniques of quality engineering developed by Dr. Genichi Taguchi. During the 1950s and 1960s, Dr. Taguchi developed a comprehensive approach to quality which touches every aspect of a product's design, manufacture, and use. In Chapter 2 we will examine Dr. Taguchi's philosophy of quality and his systems for quality in design and quality in manufacture. In later chapters we will explore his contributions to design of engineering experiments.

Statistically designed experiments are the heart of quality engineering. The days of experimentation by changing one variable at a time are over. Modern experimental designs make it possible to obtain an amazing amount of information about a process or product using a limited number of experimental runs. There is considerable debate these days about the relative merits of Taguchi's experimental designs and the designs developed by Western statisticians. This issue will be explored in later chapters. But to end the suspense early on, we should say that for the most part we are talking about the difference between a frankfurter and a hot dog—different terminology for the same basic item.

1.3 History of Quality Engineering: Japan versus U.S. Track Records

It has been said that Americans are the greatest innovators in the world. The U.S. is also recognized as a world leader in increasing productivity. But quality improvement is an area in which American industry has often come up short in international competition. Although some American companies have excellent quality in their products and processes, others have lost major market shares to foreign competition, most notably the Japanese, because of second-rate quality.

Quality systems as we understand them are a twentieth-century phenomenon. Prior to the industrial revolution, quality was built into products as they were made. It was just too expensive to make a product that was unacceptable. People involved in manufacturing knew their products inside out. Each item was produced, start to finish, either by one person or by a small team of craftspeople who knew what customers expected of the product. Large public projects such as roads, aqueducts, or palaces were built using unskilled labor, but knowledgeable supervisors and government inspectors were present at all times to make sure the job was done right. With the industrial revolution there were suddenly thousands of unskilled workers involved in high-speed manufacturing operations. For the most part they didn't understand the manufacturing processes or the technology behind them. To complicate matters further, managers had little experience running large operations. Around 1900, Frederick Taylor introduced his methods of "Scientific Management". He said that managers should manage the system and workers should do as they are told. He advocated breaking each process down to its smallest steps, and then having procedures written which described in detail how each of these steps should be carried out. Taylor's ideas were quickly adopted as a way to bring order finally to large manufacturing operations. Productivity improved, but quality remained a problem. Some argued that the way to achieve good quality was to eliminate variability: a noble, if unachievable, goal. Around 1920 Walter Shewhart introduced his control charts as a tool for stabilizing processes. He said we would never be able to remove all process variation. What we needed to do was measure the variability in production processes and remove what variability we could. His control charts provided a means of distinguishing between variability which was inherent in a process and outside disturbances to the process.

In subsequent years there were enhancements of Shewhart's charts. Dodge, Romig, Juran, and others developed tables of sampling plans to put receiving inspection and final product inspection sampling on a firm statistical foundation. But through all this the focus was on quality through inspection, or at best, quality through monitoring production processes. When World War II came along the U.S. War Department made extensive use of these statistical sampling plans when buying war materials from suppliers. They also set up training courses across the country in sampling plans, use of control charts and basic statistical concepts and methods. Through these training programs, engineers and others learned how to monitor production processes statistically.

By the end of World War II, much of the manufacturing capability in Europe and Asia had been destroyed. American manufacturing capabilities, which had grown as part of the war effort, were ready to be converted to peacetime production, and there was an insatiable demand worldwide for manufactured products. The guiding rule was that if you could make it you could sell it. If some of the items produced were defective, no problem. The customer could bring them back for repair, exchange or refund. When someone bought a new car in the 1950s, the first thing the new owner did was put a note pad and pencil in the glove compartment. As various defects were discovered, the owner made a note of them on the pad. After a couple of weeks, the car was taken back to the dealer to have the defects corrected. Again, no problem. No one expected a product as complicated as an automobile to be made completely right the first time. IBM built its reputation on quality but their early computers never worked quite right when first installed. Computer facilities all had "resident engineers" to fix the computers when they crashed. But IBM provided good service when problems arose, and that is what good quality was all about in the 1950s and 1960s—good service.

At the end of World War II, Japan's manufacturing capabilities were in much worse shape than was the case in the U.S. Japan seemed unable to produce manufactured products at a level of quality acceptable to the rest of the world. In short, "Made in Japan" meant junk. The Japanese Union of Scientists and Engineers (JUSE), took the initiative in dealing with this problem which threatened Japan's economic existence. In 1950 they asked Dr. W. Edwards Deming to come and explain his ideas for quality improvement to some of the leaders of Japanese industry. Dr. Deming had been preaching quality management in the U.S., but no one was listening—everyone was too busy making money. The Japanese made time to listen. They followed his advice. They turned their industries around. Other American quality experts also visited Japan to help them improve. Dr. Joseph Juran is credited with introducing statistical process control to Japan.

Today Japan is the acknowledged world leader in quality. Americans are now reading the works of Japanese quality experts such as Dr. Kaoru Ishikawa and Dr. Genichi Taguchi. In the 1950s, Masaaki Imai arranged tours of American plants for Japanese engineers and managers anxious to discover the secrets of American quality. Now he arranges tours of Japanese plants for Americans. In a recent book, Imai (1986) credited Japan's success in quality largely to its persistent pursuit of quality improvement in everything they do. He said that Americans depend too much on breakthroughs and technological advancements, and miss opportunities for gradual improvement.

Many American companies are making remarkable gains in product and process quality. We see some American products competing successfully in Japan. But the battle will be a long one. Today's quality requirements can't be met through inspection alone. Even SPC isn't enough. Quality must be built into products at the design stage. Dr. Juran has said, "my gloomy prognosis is that we in the West will spend the rest of the century getting fully on top of this quality crisis. Even that pace will not be realized if we continue to try to solve basic managerial problems by clever exhortations or by narrow techniques. Our managers, and especially upper managers, must arm themselves with training in how to manage for quality. They must then use that training to take active leadership of the quality function" (*Quality*, October, 1985, page 7).

1.4 Overview of Contents

The focus of this book is on the use of statistical experimental designs to economically improve product and process quality. There are many such designs available but we will consider only those which have been found most useful in engineering experiments. With these designs an experimenter can improve product and process performance, make product and process characteristics less sensitive to factors which cannot be controlled, and reduce development, manufacture, and use costs.

Chapter 2: The Taguchi approach to quality

In this chapter Dr. Taguchi's quality philosophy and his approach to quality engineering are discussed. Taguchi's definition of quality is different from those of other quality gurus, and his ideas give a fresh perspective on the whole notion of designing for quality. His system for quality engineering has several distinct steps which are discussed in Section 2.3.

Chapter 3: Two-level experiments: full factorial designs

The most useful experimental designs for product and process improvement involve sequences of trials where two different levels (high and low, fast and slow, 5% and 15%, method A and method B, etc.) are used for each controllable experimental factor. In Chapter 3, designs of this type are introduced where each possible combination of factor levels is run during the experiment. Some useful graphical tools for data analysis are also discussed. Special forms called "Response Tables" are used to simplify the calculations needed to analyze experimental data.

Chapter 4: Two-level experiments:
fractional factorial designs

Chapter 4 is the heart of this book. The designs in this chapter involve running factors at two levels each, but not every possible combination of factor levels is used. By careful selection of combinations of factor levels, the experimenter can obtain a great deal of information from only a few experimental trials. People sometimes say they can't "afford" to use statistically designed experiments, that these designs require too many experimental trials. In this chapter we find that statistical experiments are actually the best route to extracting maximum useful information in a minimum number of experimental trials.

Chapter 5: Evaluating variability using
two-level designs

Chapters 3 and 4 focus on evaluating the effect of experimental factors on average response values. Chapter 5 explains how to analyze for factor effects on variability using the experimental designs presented in Chapters 3 and 4. Dr. Taguchi was instrumental in introducing engineers to the importance of using experimental designs for this purpose. Ten years ago most experimenters thought of statistical designs as mainly a way of analyzing effects of factors on *average* values of quality characteristics. Today there is a greater awareness of reducing variability and the importance of making products and processes less sensitive to factors which affect variability. Dr. Taguchi's signal-to-noise ratio, which he recommends as a basic quality performance measure, is discussed in Chapter 5. An alternative approach based on separate measures of process average and variability is also considered.

Chapter 6: Taguchi inner and outer arrays

Chapter 6 explains Taguchi's unique approach to simultaneous analysis of the effects of controllable and uncontrollable factors. He recommends designing experiments in which these two classes of factors are initially put in separate designs, and then combining the two designs before the experiment is actually performed. Alternatives to this approach are also presented.

Chapter 7: Experimental designs for factors at three
and four levels

Sometimes it is necessary to run an experimental design at more than two levels. An experiment may involve three vendors or four possible design configurations, for example. Designs for factors run at three or four levels are described in Chapter 7.

Chapter 8: Analysis of variance in engineering design

Chapter 8 contains a somewhat nontraditional approach to a traditional statistical subject: analysis of variance. This tool appears late in this book because graphical techniques are presented earlier in the book as an easier and more informative alternative to this calculation-intense method. However, since analysis of variance is still a widely used tool in data analysis, this chapter is included to show the connection between it and the graphical approach.

Chapter 9: Computer software for experimental design

Although all the experimental designs presented in this book can be easily analyzed with paper and pencil using the special forms and graphs found in the book, frequent users may want to have the calculations performed on a computer. Five software packages which can be used with the methods in this book are discussed in Chapter 9. They represent a cross-section of the software now available for this purpose. In this chapter the reader will find information to help in deciding whether to purchase experimental design software, and what to look for in such a package.

Chapter 10: Using experiments to improve processes

How does experimental design fit into an overall plan for quality improvement? How can experimental designs be used to improve engineering design? These questions are discussed in Chapter 10.

2

The Taguchi Approach
to Quality

2.1 Definition of Quality

It can be rather discouraging to someone just beginning to explore the world of quality to discover that the leading "gurus" in quality do not agree on the definition of the word. Dr. Juran (1964) defines quality as "fitness for use." Philip Crosby, the leading promoter of the "zero defects" concept and author of *Quality Is Free* (1979), defines quality as "conformance to requirements." Dr. Deming says that "Quality should be aimed at the needs of the consumer, present and future" (1986, page 5). The American Society for Quality Control (1983) defines quality as "the totality of features and characteristics of a product or service that bear on its ability to satisfy given needs" (1983, page 4).

Although these definitions are all different, some common threads run through them:

- Quality is a measure of the extent to which customer requirements and expectations are satisfied.
- Quality is not static, since customer expectations can change.
- Quality involves developing product or service specifications and standards to meet customer needs (quality of design) and then manufacturing products or providing services which satisfy those specifications and standards (quality of conformance).

It is important to note that quality in the above context does not refer to grade or features. For example, a Chevrolet Corvette has more features and is generally considered to be a higher-grade car than a Nova. But this doesn't mean that it is of better quality. A couple with two small children may find that a four-door Nova does a much better job of meeting their requirements in terms of ease of loading and unloading, comfort when the entire family is in the car, gas mileage, maintenance, and of course, basic cost of the vehicle. As another example, whole turkeys sold in grocery stores have more water in them than did turkeys sold twenty years ago. Consumers prefer the added water because it results in juicier meat after cooking. Yet some people view turkeys with water added to them to be of poorer quality. Some people prefer quartz watches with dial faces to the less expensive watches with digital readout. Both types keep equally good time and the digital readout allows more accurate reading of time. But people pay extra for the *feature* of a dial face.

When we think of quality in terms of meeting customer expectations, we need to consider both expressed and unexpressed but understood expectations. For example, when someone buys a food product, there is the understood expectation that the product is wholesome, correctly labeled, and meets all other regulatory requirements. If someone decides to go into the pencil business and begins by surveying customers to determine what they want in wooden pencils, people might respond that they want pencils that hold their point a long time, provide clear lines, have lead which does not break easily, and erasers which last until the pencil is at least half-used. Pencil chewers would want the paint on the pencil to be lead-free. And almost everyone would want the pencils to be painted yellow, although probably no one would say so. We know people want yellow pencils because almost all pencils sold in the U.S. are yellow. This would not be an expressed requirement, but it is nevertheless important to the pencil-buying public.

Genichi Taguchi has an unusual definition for product quality: "Quality is the loss a product causes to society after being shipped, other than any losses caused by its intrinsic functions." By "loss" Taguchi refers to the following two categories:

- Loss caused by variability of function.
- Loss caused by harmful side effects.

An example of loss caused by variability of function would be an automobile that doesn't start in cold weather. The car's owner would suffer a loss if he/she had to pay someone to start the car. The car owner's employer loses the services of the employee who is now late for work. Variability in quality of services provided by airline staff can result in lost luggage, missed flights, long-distance telephone calls. An example of a loss caused by a harmful side effect would be frostbite suffered by the owner of the car which wouldn't start.

An unacceptable product which is scrapped or reworked prior to shipment is viewed by Taguchi as a cost to the company but not a quality loss. Also, losses caused when a product performs as intended are not considered quality losses by Taguchi. His reasoning here is that such situations reflect cultural or legal problems, not engineering problems. An example of this is an automobile accident caused by someone who has a radar detector and is driving too fast. The extent to which the proper functioning of the radar detector caused a loss to society is not treated as a quality loss.

In the next section we will explore further Taguchi's concept of quality from the perspective of how quality can be quantitatively measured. The material presented in the remaining chapters will help you improve quality of design no matter what definition of quality you prefer.

2.2 Loss Function

Traditionally, product quality has been measured by comparing critical product characteristics to engineering specifications for the product. Product specifications are still important, but the focus today is more on controlling process characteristics, since it is the production processes which determine the quality of the finished product. We seek to reduce process and product variability, and move quality characteristics closer to target values. This results in a reduced quality loss. Engineering specifications indicate how much variability can be tolerated in product or process characteristics. They commonly appear in forms such as the following:

1. Single value standard: Cure at 82°C.
2. Two-sided tolerance limits: Hold curing temperature between 80°C and 84°C.
3. Target value with tolerance range: 82°C ± 2°C.
4. Upper tolerance limit only: Ship product within 72 hours of receiving order.
5. Lower tolerance limit only: Bursting strength shall be at least 30 psi.

The first example, a single value standard, is not adequate for many processes. It gives no definitive criterion for deciding when the specification is not being met. We cannot hold the curing temperature at exactly 82 degrees. We could set the thermostat at 82 degrees, but we know there will be some variation around that number. How much variation is acceptable? We don't know. Dr. Deming has stressed the need for "operational definitions," that is, definitions "people can do business with." According to Dr. Deming, an operational definition consists of "(1) a criterion to be applied to an object or a group, (2) a test of the object or of the group, (3) decision: yes or no: the object or the group did or did not meet the criterion." Clearly, our first example of an engineering specification is not an operational definition; the criterion has not been defined for what constitutes acceptable variation from target; therefore a "yes or no decision" cannot be reached. The next three examples are operational definitions, assuming methods of measuring temperature and time to shipment are unambiguous. The fifth example may require clarification as to exactly how the strength test is to be performed before the specification can be treated as an operational definition.

The second example has tolerance limits but no target value. Although this provides an operational definition and adequately specifies quality requirements from the traditional engineering point of view, it isn't adequate by more recent interpretations of quality, including the idea of never-ending improvement espoused by Deming, Taguchi, Imai, and others (Deming stresses the need for quantitative measures of quality, using operational definitions, but he wants these measures to be used in reducing process variability). A classic example of the importance of having a target value as well as tolerance limits is an experience Ford Motor Company had with automobile transmissions. Ford contracted with a Japanese company to produce some transmissions for one of its automobile models. Transmissions manufactured to the same specifications in Japan and at Ford were delivered to final assembly plants and were put into automobiles basically at random. Warranty data showed that the Japanese transmissions had fewer problems. Subsequent analyses showed that all transmissions were manufactured within specifications on all critical characteristics, but the Japanese transmissions tended to be closer to target values more often. (This should not be interpreted as evidence that Ford makes inferior automobiles. Ford is a leader in the quality improvement movement in the United States. The anecdote shows Ford's willingness to share a learning experience that lesser companies would have buried in a "confidential" file.)

Examples 3 through 5 are the forms in which tolerances should be specified. Example 3 includes a specific target value. There are no central target values for examples 4 and 5, but it is clear that minimizing shipping time in example 4 and maximizing bursting strength in example 5 are "targets." Dr. Taguchi would refer to the quality characteristic in example 3 as a "nominal-is-best" type of characteristic. He would call the quality characteristic in example 4 a "smaller-is-better" type and call the characteristic in example 5 a "larger-is-better" type. We will use this terminology in section 5.5 when we discuss Taguchi's signal-to-noise ratio.

If we ignore target values and make having product characteristics within specification limits our only concern, then we are saying our loss function is of the type shown in Figure 2.1. If the characteristic is within specification, the product is acceptable and is shipped. If the characteristic is outside specification, the product is unacceptable and the problem is detected by either the manufacturer or the customer. Very little in this world is as simple as that. There are usually varying degrees of acceptability. A part can fit perfectly, fit if forced or filed, or not fit at all. A meal at a restaurant can be superb, good, adequate, below par, or awful. Ditto for customer services, ease of use, room temperature, product appearance, durability. The model of Figure 2.1 doesn't describe such situations. Dr. Taguchi suggests using a quadratic loss function of the type illustrated in Figure 2.2. This loss function clearly shows that as a characteristic moves further away from a target value, an increasing loss is incurred. In order to determine the exact form of the loss function, we would need to know the actual losses for some selected values

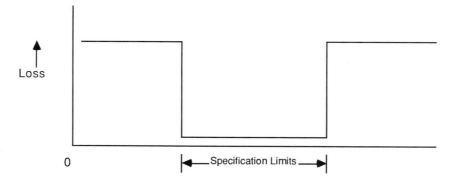

FIGURE 2.1 Traditional quality/loss function

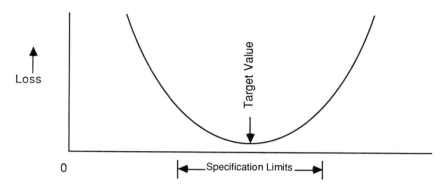

FIGURE 2.2 Quadratic quality/loss function

of the characteristic being measured. But this isn't usually a concern when designing engineering experiments. The important point is that in order to improve quality, or decrease loss, we must strive to have process and product characteristics as close to their target values as possible.

2.3 Off-Line and On-Line Quality Control

Western books on quality frequently divide quality systems into two parts: *quality of design* and *quality of conformance*. Dr. Taguchi refers to these two parts as *off-line quality control* and *on-line quality control*, respectively (see Figure 2.3).

Off-line quality control is concerned with:

1. Correctly identifying customer needs and expectations,
2. Designing a product which will meet customer expectations,
3. Designing a product which can be consistently and economically manufactured,
4. Developing clear and adequate specifications, standards, procedures and equipment for manufacture.

There are two stages in off-line quality control: *product design* stage and *process design* stage. During the product design stage a new product is developed or an existing product is modified. The goal here is to design a product which is manufacturable and will meet customer requirements. During the process design stage, production and process engineers develop manufacturing processes to meet the specifications developed during the product design stage. Activities 1, 2 and 3 are part of product design; activity 4 takes place during the process design stage. Taguchi developed a three-step approach for assuring quality within each of the two stages of off-line quality control. He called the steps *system design*, *parameter design* and *tolerance design*.

On-line quality control is concerned with manufacturing products within the specifications established during product design using the procedures de-

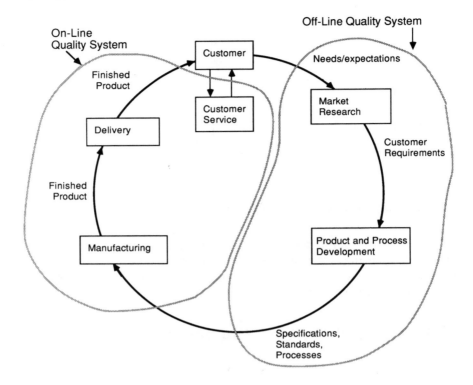

FIGURE 2.3 Production/quality system cycle

veloped during process design. Also, product and process designs may be revised if feedback from customers reveals opportunities for improvement. Taguchi (1986) identifies two stages of on-line quality control. Stage 1, *production quality control methods,* has three forms: *process diagnosis and adjustment, prediction and correction,* and *measurement and action.* Stage 2 is *customer relations.*

In the remainder of this section we will briefly go through the steps in Dr. Taguchi's approach to quality. Figure 2.4 summarizes these steps.

Product design (off-line QC, stage 1)

1. *System design:* applying engineering and scientific knowledge to develop a prototype design which meets customer requirements. Initial selection of parts, materials and manufacturing technology are made at this time. The emphasis here is on using the best available technology to meet customer requirements *at lowest cost.* A key difference between this step in Taguchi's approach and the prototype design step in many Western R&D departments is Taguchi's focus on proven technology, low-cost parts, and customer requirements rather than on using the latest technology and exotic or expensive parts.

2. *Parameter design:* determination of optimal setting for product parameters. The goal here is to minimize manufacturing and product lifetime costs

		Concerns:	QA Steps:
OFF-LINE QUALITY CONTROL	**Stage 1:** **PRODUCT** **DESIGN**	1. Identify customer needs and expectations 2. Design a product to meet customer needs and expectations 3. Design a product which can be consistently and economically manufactured	**QA Steps:** 1. System Design 2. Parameter Design 3. Tolerance Design
	Stage 2: **PROCESS** **DESIGN**	**Concerns:** 1. Develop clear and adequate specification standards, procedures and equipment for manufacture	**QA Steps:** 1. System Design 2. Parameter Design 3. Tolerance Design
ON-LINE QUALITY CONTROL	**Stage 1:** **PRODUCTION**	**Concerns:** 1. Manufacture products within specifications established during product design using procedures developed during Process Design	**Form 1:** Process Diagnosis and Adjustment **Form 2:** Prediction and Correction **Form 3:** Measurement and Action
	Stage 2: **CUSTOMER** **RELATIONS**	**Concerns:** 1. Provide service to customers and use information on field problems to improve product and manufacturing process designs	**Actions:** 1. Repair, replacement or refund 2. Feed back information on field problems 3. Change product and process specifications/design

FIGURE 2.4 Taguchi's quality system

by minimizing performance variation. This involves making the product design *robust*—insensitive to noise factors. A *noise factor* is an uncontrollable source of variation in the functional characteristics of a product. Taguchi identifies three types of noise factors: *external noise*, or variation in environmental conditions, such as dust, temperature, humidity, or supply voltages; *internal noise,* or deterioration, such as product wear, material aging, or other changes in components or materials with time or use; and *unit-to-unit noise*, which is differences in products built to the same specifications caused by variability in materials, manufacturing equipment, and assembly processes. The parameter design step involves use of experimental designs to determine the impact of both controllable and uncontrollable (noise) factors on product characteristics. The goal is to set controllable factors at levels which will make the product robust with respect to noise factors. Examples of robust design include an automobile part which can withstand shock and vibration (external noise), or a food product with a long shelf life (internal noise), or a replacement part that will fit properly (unit-to-unit noise).

3. *Tolerance design*: establish tolerances around the target (nominal) values established during parameter design. The goal is to set tolerances wide (to reduce manufacturing costs) while still keeping the product's functional characteristics within specified bounds.

Process design (off-line QC, stage 2)

1. *System design*: select the manufacturing process on the basis of knowledge of the product and current manufacturing technology. The focus here is on building to specification using existing machinery and processes whenever possible.

2. *Parameter design*: determine appropriate levels for controllable production process parameters. The goal here is to make the process robust—to minimize the effects of noise on the production process and finished product. Experimental designs are used during this step.

3. *Tolerance design*: establish tolerances for the process parameters identified as critical during process parameter design. If the product or process parameter design steps are poorly done, it may be necessary here to tighten tolerances or specify higher-cost materials or better equipment, thus driving up manufacturing costs.

Production quality control methods (on-line QC, stage 1)

Dr. Taguchi identifies three forms of on-line quality control:

1. *Process diagnosis and adjustment*: the process is monitored at regular intervals; adjustments and corrections are made as needed.
2. *Prediction and correction*: a quantitative process parameter is measured at regular intervals. The data are used to project trends in the process. Whenever the process is found to be too far off target, the process is adjusted to

correct the situation. This method is also called *feedback* or *feedforward* *control*.

3. *Measurement and action*: quality by inspection. Every manufactured unit is inspected. Defective units are reworked or scrapped. This is the most expensive and least desirable form of production quality control since it doesn't prevent defects from occurring or even identify all defective units.

Customer relations (on-line QC, stage 2)

Customer service can involve repair or replacement of defective products, or compensation for losses. The complaint handling process should be more than a customer relations operation. Information on types of complaints and failures, and customer perceptions of products, should be promptly fed back to relevant functions within the organization for corrective action.

Figure 2.5 summarizes at which point in the product design-manufacture-delivery cycle it is possible to deal with the three types of noise. In this figure, the "product design" stage is called "R&D" and the "process design" stage is called "production engineering." Figure 2.5 shows that noise factors can be dealt with during several different steps in the process. However, external and internal noise can be reduced most effectively and economically at the parameter design step of product design, and unit-to-unit noise is best handled during the parameter design steps of product design and process design.

Department Countermeasure			Type of Noise		
			External	Internal	Unit-to-Unit
Off-line quality control	R & D	(1) System design (2) Parameter design (3) Tolerance design	Δ Δ •	Δ Δ Δ	Δ Δ Δ
	Production Engin-eering	(1) System design (2) Parameter design (3) Tolerance design	◊ ◊ ◊	◊ ◊ ◊	Δ Δ Δ
On-line quality control	Production	(1) Process diagnosis and adjustment (2) Prediction and correction (3) Measurement and action	◊ ◊ ◊	◊ ◊ ◊	Δ Δ Δ
	Customer relations	After-sales service	◊	◊	◊

Δ Possible
• Possible, but should be a last resort
◊ Impossible

FIGURE 2.5 Dealing with noise factors (From *Introduction to Quality Engineering,* G. Taguchi; UNIPUB, 1986, with permission)

2.4 Taguchi's Quality Philosophy

Genichi Taguchi's impact on the world quality scene has been far-reaching. His quality engineering system, outlined in the previous section, has been used successfully by many companies in Japan and elsewhere. He stresses the importance of designing quality into products and processes, rather than depending on the more traditional tools of on-line quality control. Taguchi's approach differs from that of other leading quality gurus in that he focuses more on the engineering aspects of quality rather than on management philosophy or statistics. Also, Dr. Taguchi uses experimental designs primarily as a tool to make products more *robust*—to make them less sensitive to noise factors. That is, he views experimental design as a tool for reducing the effects of variation on product and process quality characteristics. Earlier applications of experimental design focused more on optimizing average product performance characteristics without considering effects on variation. Kackar (1986) summarizes Taguchi's quality philosophy into the following seven basic elements (Material in italics is quoted directly from that paper):

(1) *An important dimension of the quality of a manufactured product is the total loss generated by that product to society.* At a time when the "bottom line" appears to be the driving force for so many organizations, it seems strange to see "loss . . . to society" as part of product quality. It is an interesting concept, and not necessarily part of Western capitalistic thinking, but then again, the U.S. approach hasn't done much for the balance of trade. Incidentally, the experimental designs presented in this book are useful to engineers and scientists regardless of one's definition of "quality."

(2) *In a competitive economy, continuous quality improvement and cost reduction are necessary for staying in business.* This is a hard lesson to learn. Masaaki Imai (1986) argues very persuasively that the principal difference between Japanese and American management is that American companies look to new technologies and innovation as the major route to improvement, while Japanese companies focus more on *"Kaizen,"* gradual improvement in everything they do. Taguchi stresses use of experimental designs in parameter design as a way of reducing quality costs. He identifies three types of quality costs: R&D costs, manufacturing costs, and operating costs. All three costs can be reduced through use of appropriate experimental designs.

(3) *A continuous quality improvement program includes incessant reduction in the variation of product performance characteristics about their target values.* Again *Kaizen*. But with a focus on product and process variability. This does not fit the mold of quality being conformance to specification. Meeting specs must be viewed as a first step in quality, not a final goal.

(4) *The customer's loss due to a product's performance variation is often approximately proportional to the square of the deviation of the performance characteristic from its target value.* This concept of a quadratic loss function, illustrated in Figure 2.2, says that any deviation from target results in some loss to the customer, but that large deviations from target result in severe losses.

(5) *The final quality and cost of a manufactured product are determined*

to a large extent by the engineering designs of the product and its manufacturing process. This is so simple, and so true. Much of what we call quality control today is "problem solving" which tries to resolve on-line the symptoms of chronic problems created during product or process design. The future belongs to companies which, once they understand the variabilities of their manufacturing processes using statistical process control, move their quality improvement efforts upstream to product and process design.

(6) *A product's (or process') performance variation can be reduced by exploiting the nonlinear effects of the product (or process) parameters on the performance characteristics.* This is an important statement because it gets to the heart of off-line QC. Instead of trying to tighten specifications beyond a process' capability, perhaps a change in design can allow specifications to be loosened. As an example, suppose that in a heating process the tolerance on temperature is a function of the heating time in the oven. The tolerance relationship is represented by the band in Figure 2.6. For example, if a process specification says the heating process is to last 4.5 minutes, then the temperature must be held between 354.0 and 355.0 degrees, a tolerance interval 1.0 degrees wide. Perhaps the oven cannot hold this tight a tolerance. One solution would be to spend a lot of money on a new oven or new controls. Another possibility would be to change the time for the heating process to, say, 3.75 minutes. Then the temperature would need to be held to between 358.0 and 360.6 degrees, an interval of width 2.6 degrees. If the oven could hold this tolerance, the most economical decision might be to adjust the specifications. This would make the process less sensitive to variation in oven temperature.

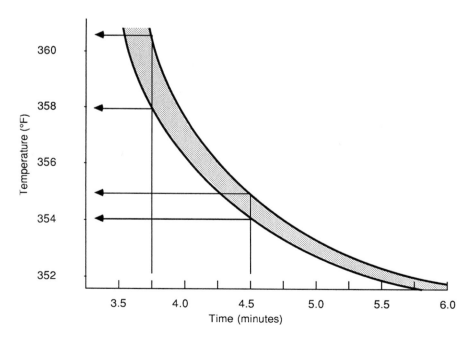

FIGURE 2.6 Time-temperature relationship

(7) *Statistically designed experiments can be used to identify the settings of product (and process) parameters that reduce performance variation.* And hence improve quality, productivity, performance, reliability, and profits. Statistically designed experiments will be the strategic quality weapon of the 1990s.

3

Two-level Experiments: Full Factorial Designs

3.1 Experimentation as a Learning Process

We have all performed experiments. In high school we dipped litmus paper in an acid and watched it change color; we measured the force needed to move a block across a table; we searched for the reproductive organs in a worm. At home we "experimented" with different explanations for arriving home late on Saturday night. We also experimented with different stances, grips, and bats while playing baseball. The experiments that we are going to study in this book are intended to help improve the quality of designs and processes. By "improve" we mean:

- *Optimizing the average response value.* This has traditionally been the main focus of experimental designs as used in the West. We want to maximize yield, have tube outside diameters be 1.2 cm, or have polyurethane sheets be as smooth as possible. Experimental designs were used to identify which combinations of settings, or "levels," for certain key factors produced the best average value for the product or process characteristic of interest. Chapters 3 and 4 will focus on this use of experimental design because it is important and it is the easiest application to understand.
- *Minimizing effects of variability on process or product performance.* This is also called *robust design*. In the Taguchi approach to quality engineering, the primary role of experimental design is to make the process or product insensitive (robust) to variation in uncontrolled factors. Put another way, robust design refers to reducing the variability in some performance measure by making the measure insensitive to noise factors. Taguchi recommends that this be done during the "parameter design" phase of off-line quality control (see section 2.3). This use of experimental designs will be explained in Chapters 5 and 6.

Before we look at some examples of experiments, a few terms need to be defined. An *experiment* can be defined broadly as any act of observing. So, measuring the breaking strength of thread or the electrical conductivity of an alloy, interviewing voters, or just watching clouds move across the sky on a summer day could all be called experiments. We will use a more restrictive definition, however, and say that an experiment is a series of trials or tests which produce quantifiable outcomes. Observing the temperature at your front door at sunrise tomorrow would be an experiment. Some people call it a *random* experiment since its observed value cannot be predicted exactly in advance. An experiment where the outcome can be completely predicted in advance is then called a *deterministic* experiment. Although experiments produce quantifiable outcomes, the initial data need not be numbers. It is enough that the information be convertible to numbers. For example, we could observe whether a piece of equipment works or not, and then say the outcome of the experiment was "1" if it worked and "0" if it did not. If you evaluated the taste of your first cup of coffee in the morning as good, fair, or poor, you could then arbitrarily assign values 1, 2, and 3, respectively, to these three categories.

Experiments are carried out for a variety of reasons. Industrial experiments are generally performed to explore, estimate, or confirm.

- *Exploration*: gather data to learn more about a process or product characteristic. Suppose, for example, we want to better understand the effects of curing time and temperature on the strength of a molded part. We will cure specimens of the product at several different temperatures and for different times and then measure their breaking strengths.
- *Estimation*: use data to estimate the effects of certain variables on other variables. For example, if our exploratory study showed that the breaking strength of a molded part is affected by both curing time and temperature, we might want to estimate the average breaking strengths at various combinations of these two process factors. This information could be used to estimate the settings of the two factors which would maximize breaking strength.
- *Confirmation*: gather data to verify a hypothesis about a relationship among variables. For example, once the "optimal" curing time and temperature have been determined, additional experiments are run at and near these values to verify that the settings are in fact "best."

Example A

Buy two different brands of chewing gum. Chew several packs of each (one stick at a time). Which brand is more enjoyable to chew? Which brands hold their flavor longer? (*Exploration*)

For each stick of gum chewed, record how long (in minutes) the flavor is clearly detectable. Estimate the average time the flavor lasts for each brand. (*Estimation*)

Brand A appears to have longer lasting flavor than brand B, based on an exploratory study. Data are collected on the length of time flavor lasts for thirty sticks to verify this conclusion. (*Confirmation*)

Example B

An industrial equipment supply company has been receiving complaints lately from some of its customers that orders are not being delivered on time. An investigation of current delivery performance shows that, on the average, orders are being shipped within four working days. But there is considerable variability in delivery, and many orders take much longer to be shipped. (*Exploration*)

Of the orders selected for the study, 62 percent are shipped in four days, but 14 percent take at least seven days to be shipped. (*Estimation*)

In an effort to reduce variability in delivery times, changes are made in scheduling. The new procedures seem to be effective. In a follow-up study, it is found that among 400 orders checked, 96 percent were shipped within six days. (*Confirmation*)

In this chapter we will learn how to use experiments involving two to four experimental factors. The designs can be very helpful to the engineer who is trying to improve the performance or robustness of a product or process. This chapter also forms a basis for our discussions in the next chapter, in which use of designs involving up to fifteen factors will be explained.

3.2 Traditional Scientific Experiments

"I'm an experienced engineer. I know how to test products. The right way to do it is to vary one factor at a time, holding the others fixed. That way I can quickly get accurate information on performance. Right?" Not always. Let's begin with an example.

Suppose the strength of a part is a function of three factors: A, B, and C. We want to evaluate the strength of the part for two different values of each of the three factors so that levels can be selected to maximize strength. Using the traditional approach of varying one factor at a time, we might perform four tests at the levels indicated in Figure 3.1, where *1* and *2* represent the two levels at which each factor is to be set during the experiment. Generally, level *1* is called the "low level" for the factor and *2* the "high level."

Suppose we perform the four trials and obtain the results listed in Figure 3.2. Based on this information, we would conclude that strength of the part is maximized (at 130) by setting factors A and C at level 1 and factor B at level 2. Figure 3.2 also suggests that changing factor A from level 1 to level 2

Trial	LEVELS OF FACTORS			Response (strength)
	A	*B*	*C*	
1	1	1	1	y_1
2	2	1	1	y_2
3	1	2	1	y_3
4	1	1	2	y_4

FIGURE 3.1 Factor settings for experimental design

Trial	LEVELS OF FACTORS			Strength
	A	*B*	*C*	
1	1	1	1	125
2	2	1	1	100
3	1	2	1	130
4	1	1	2	105

FIGURE 3.2 Observed strengths for example experiment

while holding factors B and C at level 1 decreases part strength by $125 - 100 = 25$. Similarly, changing factor B from level 1 to level 2 increases part strength by $130 - 125 = 5$, and changing factor C from level 1 to level 2 decreases part strength by $125 - 105 = 20$.

But, suppose factors A and B *interact* with each other. That is, suppose these two factors do not have independent effects on strength, but rather the effect of one factor on strength is affected by the level of the other factor. Suppose that because of this interaction, A and B combined contribute to part strength as listed in Figure 3.3 and illustrated in Figure 3.4. Also, suppose Factor C contributes strength of 70 at its level 1 and strength equal to 50 at level 2. Then, the strength values for the four experimental trials would be as calculated in Figure 3.5.

Based on the strength values in Figure 3.5, we would again decide that total part strength was maximized (at 130) by setting factors A, B, and C at levels 1, 2, and 1, respectively. But this is incorrect. In order to maximize strength we should set factors A, B, and C at levels 2, 2, and 1, respectively, giving a total strength of $90 + 70 = 160$.

How do we avoid falling into traps of the type illustrated above? By using appropriate statistically designed experiments.

3.3 Three-factor Design

The experimental design considered in this section can be used to evaluate the effects of three different factors, where each factor is set at two different levels. All possible combinations of levels are included, so there are

LEVEL		Strength contribution
A	*B*	*of factors A and B*
1	1	55
1	2	60
2	1	30
2	2	90

FIGURE 3.3 Strength values with interaction effect

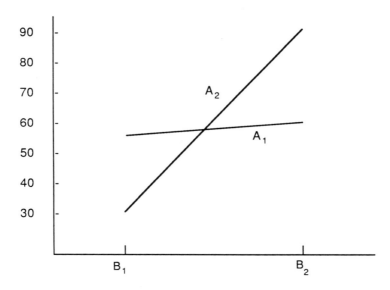

FIGURE 3.4 Graph of strength values with interaction effect

$2 \times 2 \times 2 = 2^3 = 8$ trials in the experiment. The two levels used for a factor may represent two selected values of a continuous factor (temperature, pressure, time, thickness, etc.) or two different discrete possibilities (machine A or B, operator Amanda or Sue, new method or old method, water in cooling sleeve turned on or off, acid dip prior to applying coating or no acid dip, two or three coats of finish). Although these controllable factors can be continuous or discrete, the process characteristic which is a function of these factors and which will be observed during our experiment (called the *response*) should be a continuous variable.

The eight trials which comprise the design are listed in Figure 3.6. We will study this design rather closely since many important characteristics of good experimental designs can be illustrated with it.

According to Figure 3.6, the first trial would be run with all factors at their *low* levels. In the second trial, factors A and B would be set at their *low* levels and factor C at its *high* level, etc. The designation of *low* and *high* is arbitrary and need not imply that the low level has a numerically lower value than the high level.

	LEVELS OF FACTORS			
Trial	*A*	*B*	*C*	*Strength*
1	1	1	1	$55 + 70 = 125$
2	2	1	1	$30 + 70 = 100$
3	1	2	1	$60 + 70 = 130$
4	1	1	2	$55 + 50 = 105$

FIGURE 3.5 Strengths for trials with factor interactions

Trial	LEVELS OF FACTORS			
number	*A*	*B*	*C*	*Response*
1	1	1	1	y_1
2	1	1	2	y_2
3	1	2	1	y_3
4	1	2	2	y_4
5	2	1	1	y_5
6	2	1	2	y_6
7	2	2	1	y_7
8	2	2	2	y_8

FIGURE 3.6 Design matrix for a three-factor, eight-run experiment

The experimental design described in Figure 3.6 has the property of being *orthogonal.* This is an important property which will be discussed later. Briefly, orthogonality allows estimation of the average effects of factors without fear that the results are being distorted by effects of other factors. In this book it will generally be pointed out what factors in a design are orthogonal and what factors are not. But there is also an easy trick to determine orthogonality: First, replace the 1s and 2s in the design matrix by -1s and 1s, respectively. (This is, incidentally, the notation preferred by many Western statisticians.) Then, multiply together the corresponding row values from each of two columns. Finally, add up these products. If the sum is equal to zero, the columns are orthogonal and the effects represented by these columns are also said to be orthogonal.

Example

We will verify that columns A and B of the design matrix in Figure 3.6 are orthogonal. If we replace 1 by -1 and 2 by 1, Figure 3.6 becomes Figure 3.7.

Trial	LEVELS OF FACTORS			
number	*A*	*B*	*C*	*Response*
1	-1	-1	-1	y_1
2	-1	-1	1	y_2
3	-1	1	-1	y_3
4	-1	1	1	y_4
5	1	-1	-1	y_5
6	1	-1	1	y_6
7	1	1	-1	y_7
8	1	1	1	y_8

FIGURE 3.7 Design matrix for a three-factor, eight-run experiment using alternate $(-1,1)$ notation

If the corresponding row values for factors A and B are multiplied together and the products added, the sum is:

$$
\begin{aligned}
\text{Sum} &= (-1)\,(-1)+(-1)\,(-1)+(-1)\,(1)+(-1)\,(1) \\
&\quad +(1)\,(-1)+(1)\,(-1)+(1)\,(1)+(1)\,(1) \\
&= 1+1-1-1-1-1+1+1 \\
&= 0.
\end{aligned}
$$

The reader might want to verify that columns A and C are orthogonal and also that columns B and C are orthogonal.

The *effect* of a factor on a response variable is the change in the response when the factor goes from its low level to its high level. With the design in Figure 3.6, we can estimate the effect of each factor by finding the average value for the response variable at the high level of the factor, and also at the low level of the factor, and then taking the arithmetic difference between these two average values. Since the columns for the factors in the design are orthogonal, our estimate of the effect of any factor on the response will not be distorted by the effects of other factors.

In order to estimate the effect of factor A on the average value of the response variable, we first add together the four observed responses at level 1 of factor A. We then divide the sum by 4 to obtain the average response at *low* level of factor A. This average is denoted by $\overline{A}_1$. In a similar manner we obtain the average response at the *high* level of A, and denote it by $\overline{A}_2$. Then, the *effect* of A on the average response is equal to the average response at the high level of A minus the average response at the low level of A. That is, if y_i is the value of the response variable on the *ith* trial as listed in Figure 3.6, then the estimated effect of factor A is:

$$
\begin{aligned}
\text{Effect of A} &= (\text{average at "2" value of A}) \\
&\quad - (\text{average at "1" value of A}) \\
&= \overline{A}_2 - \overline{A}_1 \\
&= \frac{(y_5+y_6+y_7+y_8)}{4} - \frac{(y_1+y_2+y_3+y_4)}{4}
\end{aligned}
$$

The effects of factors B and C on the response variable can be similarly estimated:

$$
\begin{aligned}
\text{Effect of B} &= (\text{average at "2" value of B}) \\
&\quad - (\text{average at "1" value of B}) \\
\text{Effect of C} &= (\text{average at "2" value of C}) \\
&\quad - (\text{average at "1" value of C})
\end{aligned}
$$

Figure 3.8 is an illustration of a *response table*,* which is used to calculate estimated effects. In the first column the order in which the trials are to be run should be listed. We will discuss this shortly. The second column

*We are indebted to Mr. Randall F. Culp, Manager of Advanced Technology and Reliability Projects at General Electric's Medical Division, for suggesting the concept of the response table to us.

Random Order Trial Number	Standard Order Trial Number	Response Observed Values y	A 1	A 2	B 1	B 2	C 1	C 2
____	1	y_1	y_1		y_1		y_1	
____	2	y_2	y_2		y_2			y_2
____	3	y_3	y_3			y_3	y_3	
____	4	y_4	y_4			y_4		y_4
____	5	y_5		y_5	y_5		y_5	
____	6	y_6		y_6	y_6			y_6
____	7	y_7		y_7		y_7	y_7	
____	8	y_8		y_8		y_8		y_8
TOTAL		(sum of measurements in columns above goes here)						
NUMBER OF VALUES		8	4	4	4	4	4	4
AVERAGE		$\overline{y}$	$\overline{A}_1$	$\overline{A}_2$	$\overline{B}_1$	$\overline{B}_2$	$\overline{C}_1$	$\overline{C}_2$
EFFECT			$\overline{A}_2 - \overline{A}_1$		$\overline{B}_2 - \overline{B}_1$		$\overline{C}_2 - \overline{C}_1$	

FIGURE 3.8 Response table for a three-factor experiment

gives the trial number as listed in the design matrix in Figure 3.6. The third column represents the observed response values. The remaining columns show which response values should be used when calculating the various averages. For example, in the fourth column we see that y_1, y_2, y_3, and y_4 should be used when calculating $\overline{A}_1$.

Example 3.1

The yield of a chemical reaction was thought to be a function of three variables:

Formulation (F)

Mixer speed (S)

Temperature (T)

An eight-run, two-level experiment was conducted based on these three factors. The levels selected for the factors are given in Figure 3.9.

Based on Figure 3.6, the eight experimental runs for our example are as listed in columns 3, 4, and 5 of Figures 3.10 (coded) and 3.11 (uncoded).

Level	Formulation	Mixer speed	Temperature
1	A	60 rpm	70°C
2	B	80 rpm	82°C

FIGURE 3.9 Levels selected for factors

Random Order Trial Number	Standard Order Trial Number	Formulation F	Mixer Speed S	Temperature T	Yield Y
____	1	1	1	1	____
____	2	1	1	2	____
____	3	1	2	1	____
____	4	1	2	2	____
____	5	2	1	1	____
____	6	2	1	2	____
____	7	2	2	1	____
____	8	2	2	2	____

FIGURE 3.10 Coded design matrix for example

Random Order Trial Number	Standard Order Trial Number	Formulation F	Mixer Speed S	Temperature T	Yield Y
4	1	A	60	70	_____
1	2	A	60	82	_____
8	3	A	80	70	_____
5	4	A	80	82	_____
6	5	B	60	70	_____
3	6	B	60	82	_____
2	7	B	80	70	_____
7	8	B	80	82	_____

FIGURE 3.11 Experimental design for example

Before the experiment is performed, the order in which the trials will be run should be randomly assigned. This can be done in several ways. Two easy possibilities are:

- Write the numbers 1 through 8 on eight separate slips of paper, put the slips in a bowl, mix them up, and take them out one at a time. Run the eight experimental trials in the order in which the slips were drawn.
- Use the Table of Random Orderings for Eight-Run Experiments in the appendix of this book.

You could also use an eight-sided die from your son's or daughter's Dungeons & Dragons™ game, and record the order in which you first get a 1, 2, . . . , 8. The assignment of order of runs in column 1 of Figure 3.11 was obtained from the beginning of the Table of Random Orderings for Eight-Run Experiments in the appendix.

It is important for the validity of the experiment that the trials be run in random order. If the trials are run in the order listed in Figure 3.10, say, then unsuspected factors which change with time may distort the analysis and result in misleading conclusions. For example, suppose one of the chemicals used in the experiment deteriorated over the course of the experiment. The yield might be lower than expected later in the experiment. Since formulation A was used during the first four trials (using standard order) and formulation B used during the second four trials, this reduction in yield might be mistakenly attributed to formulation effect. Randomization minimizes the chance of this sort of thing happening.

To simplify the job of collecting data, and to minimize the chance of error, the person or group assigned to perform the experiment and record the results should be given a listing of the experimental trials *in the order in which they are to be performed.* Figure 3.12 is a *report form* which meets this need. It contains the same information provided in Figure 3.11, but now the trials are listed in the (random) order in which they are to be performed, rather than in standard order. Note also that the observed values for the response can be recorded directly on the report form. The observed responses for the example we are considering have been added to the report form in Figure 3.13.

To simplify calculating factor effects, the observed values of the response variable should be transferred from the report form to a response table. Figure 3.14 is a blank response table for an eight-run, two-level experiment in

Run Order For Experiment	Standard Order Trial Number	Formulation F	Mixer Speed S	Temperature T	Yield Y
1	2	A	60	82	_____
2	7	B	80	70	_____
3	6	B	60	82	_____
4	1	A	60	70	_____
5	4	A	80	82	_____
6	5	B	60	70	_____
7	8	B	80	82	_____
8	3	A	80	70	_____

FIGURE 3.12 Report form for operator

Run Order For Experiment	Standard Order Trial Number	Formulation F	Mixer Speed S	Temperature T	Yield Y
1	2	A	60	82	166
2	7	B	80	70	179
3	6	B	60	82	187
4	1	A	60	70	164
5	4	A	80	82	160
6	5	B	60	70	184
7	8	B	80	82	182
8	3	A	80	70	161

FIGURE 3.13 Completed report form

three factors. In Figure 3.15 the run order numbers and yield data from Figure 3.13 have been added to the form. Figure 3.16 is a completed response table. Note at the bottom of Figure 3.16 that the average response at the low and high levels of each factor, and resulting factor effects, are calculated. For example, the response averages at the low and high levels of factor F are 651/4 = 162.8 and 732/4 = 183.0, respectively. The F factor effect is then 183.0 − 162.8 = 20.2. Since the effect for F is greater than zero, the average response is higher for the high (B) level of F than for the low (A) level. For S, on the other hand, the estimated effect is less than zero, indicating that the average response is higher at the low level of S than at the high level. The effect for factor T is so small that it should probably be viewed as due to random variation rather than a "real" temperature effect. The *grand average*, $\bar{y}$, is calculated to be 1383/8 = 172.9 at the bottom of column 3 of Figure 3.16.

Graphical display of factor effects

The effects of factors used in an experiment can be presented on a graph using the following procedure:

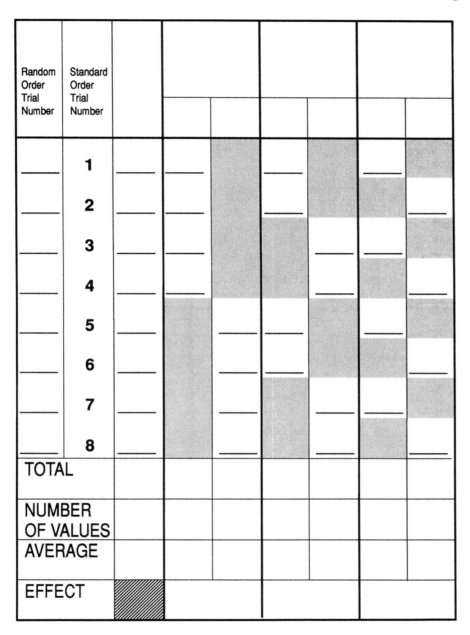

FIGURE 3.14 Blank response table

Random Order Trial Number	Standard Order Trial Number	Response Observed Values Y	Formulation F		Mixer Speed S		Temperature T	
			1 A	2 B	1 60	2 80	1 70	2 82
4	1	164	___		___		___	
1	2	166	___		___			___
8	3	161	___			___	___	
5	4	160	___			___		___
6	5	184		___	___		___	
3	6	187		___	___			___
2	7	179		___		___	___	
7	8	182		___		___		___
TOTAL								
NUMBER OF VALUES	8							
AVERAGE								
EFFECT								

FIGURE 3.15 Report table form with experimental data added

1. Identify the largest and smallest average responses. (The average responses are found in the "AVERAGE" row of the response table.)
2. Draw a vertical scale to include all of these average values.
3. Draw a horizontal line at the grand average value (last number in column 3 of the response table).
4. For each factor, plot the average response value at the high level and also at the low level. Plot one point directly over the other. One point will be above the grand average line and the other will be below. Also, they will be equidistant from the grand average line.
5. Label the points and connect each pair of points by a vertical line.

Random Order Trial Number	Standard Order Trial Number	Response Observed Values y	Formulation F		Mixer Speed S		Temperature T	
			1	2	1	2	1	2
4	1	164	164		164		164	
1	2	166	166		166			166
8	3	161	161			161	161	
5	4	160	160			160		160
6	5	184		184	184		184	
3	6	187		187	187			187
2	7	179		179		179	179	
7	8	182		182		182		182
TOTAL		1383	651	732	701	682	688	695
NUMBER OF VALUES		8	4	4	4	4	4	4
AVERAGE		172.9	162.8	183.0	175.3	170.5	172.0	173.8
EFFECT				20.2		−4.8		1.8

FIGURE 3.16 Completed response table

This graphical technique has been used by Kackar and Shoemaker (1986) and others. Ott (1975) used similar graphs, but he drew his effect lines at an angle rather than vertically, so the point for the "low" level of a factor was slightly to the left of the point for the "high" level.

The factor effects for the illustrative example we have been considering are plotted in Figure 3.17. Note that the larger the vertical line, the larger the change in response when going from level 1 to level 2 of a factor. In Chapter 8 an important statistical tool called Analysis of Variance (ANOVA) will be discussed. It will be pointed out there that the statistical significance of a factor is directly related to the lengths of the vertical lines in Figure 3.17.

A graphical display such as Figure 3.17 can be used, in conjunction with

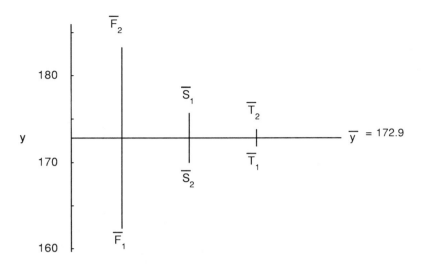

FIGURE 3.17 Graphical presentation of factor effects

a response table, to identify appropriate settings for factors when the experimenter wants to *maximize* ("larger-is-better") or *minimize* ("smaller-is-better") the average response.

In order to *maximize* the average response for our example, we should, based on Figure 3.17, set factors F and S at levels 2 and 1, respectively, since average responses are higher at these levels of these factors than at the other levels of the two factors. An experimenter might decide, based on Figure 3.17, that the contribution of factor T is negligible. In this case, T would not be considered further in the analysis. The "contribution" of factor F at its high level is the amount by which the average response at level 2 for F differs from the average response at the grand average. That is, it is equal to $\bar{F}_2 - \bar{y}$. At factor levels F_1 and S_1 the average response is estimated to be:

$$
\begin{aligned}
\bar{y}_{max} &= \bar{y} + (\bar{F}_2 - \bar{y}) + (\bar{S}_1 - \bar{y}) \\
&= 172.9 + (183.0 - 172.9) + (175.3 - 172.9) \\
&= 172.9 + 10.1 + 2.4 \\
&= 185.4
\end{aligned}
$$

Similarly the *minimum* average response would be expected when F and S are at levels F_2 and S_2. Then the average response is estimated to be:

$$
\begin{aligned}
\bar{y}_{min} &= \bar{y} + (\bar{F}_1 - \bar{y}) + (\bar{S}_2 - \bar{y}) \\
&= 172.9 + (162.8 - 172.9) + (170.5 - 172.9) \\
&= 172.9 - 10.1 - 2.4 \\
&= 160.4
\end{aligned}
$$

3.4 Replicating Experiments

In the experimental design in the previous section, each possible combination of three factors was used exactly once. With an orthogonal design, each factor's effect can be estimated without being biased by the other factors. However, if there is excessive variability in the process being studied, the estimates might be far from the true average values. It is much like trying to estimate the fairness of a coin by tossing it twice, or estimating the gas mileage of a car based on one tank of gasoline. The effects of high variability on experimental results can often be reduced by performing an experiment more than once—that is, by *replicating* the experiment. There are at least four benefits of replicating experiments:

1. Average values have less variability than individual measurements, so our calculated averages will tend to be closer to "true" factor effects.
2. Without replications, a single erroneous or unusual sample value can distort the whole analysis.
3. Data from replicated experiments can be used to estimate the amount of variability in the process. This important topic will be discussed further in Chapter 5.
4. Data from replicated experiments can be used to determine which factors affect the mean level of the process and which factors affect the variability of the process. This will also be discussed in Chapter 5.

Each complete set of the eight factor combinations in the experimental design is called a *replication*. Thus, if the complete experiment was carried out three times (a total of twenty-four trials), we would say the experiment was replicated three times. The order in which the twenty-four trials are performed should be randomized. Sometimes it is appropriate to perform experiments one replication at a time. For example, if an eight-run experiment is to be replicated three times, each set of eight trials is run as a separate group, with randomization of order *within each group of eight trials*. This is called *blocking*, and will be discussed further in section 4.5. When analyzing data from a replicated experiment, we first calculate the average response at each combination of factor levels. These eight averages are then treated the same as the eight individual responses obtained in the last section from an unreplicated experiment.

A key goal in quality improvement is to reduce process and product variability. Put another way, we seek to reduce the impact of noise factors on critical quality characteristics. If the amount of variability is estimated at each point in the design, it is possible to determine which combination of factors minimize variability in the process or product. In Chapter 5 we will see how to use experimental designs to reduce process/product variability. In Chapter 6 we will discuss an approach of Dr. Taguchi's in which levels of certain noise factors are controlled and changed in a systematic way during the experiment.

Example 3.2

In section 3.3 we analyzed an unreplicated three-factor experiment. That experiment had eight trials. Suppose now that the experiment was replicated twice. That is, each of the eight combinations of factor levels was run twice,

Standard Order Trial Number	Formulation F	Mixer Speed S	Temperature T	Yield (Y)		Sum of Repli- cations
				Replicate No. 1	Replicate No. 2	
1	A	60	70	164	160	324
2	A	60	82	166	168	334
3	A	80	70	161	163	324
4	A	80	82	160	157	317
5	B	60	70	184	182	366
6	B	60	82	187	184	371
7	B	80	70	179	181	360
8	B	80	82	182	180	362

FIGURE 3.18 Example with two replicates

for a total of sixteen trials. The results of the experiment are reported in Figure 3.18. The 16 trials were run in random order, but the information and data are listed in standard order in the figure. Values in the "Sum of Replications" column of Figure 3.18 are obtained by adding together the values of the two yields in the same row. The information from this experiment is recorded in the response table in Figure 3.19. Note that since each yield value in Figure 3.19 is the sum of two observed yields, the numbers in the "Number of Values" row are doubled in comparison to those found in Figure 3.16. The factor effects are displayed graphically in Figure 3.20. For this example the results of the replicated experiment are basically the same as those of the unreplicated experiment in section 3.3.

3.5 Factor Interactions

The last two sections explained how to measure the effects of three separate factors using an eight-run experimental design. The three factors were treated as being *additive*. That is, each factor makes its contribution to the response variable independent of the other two factors. But sometimes the factors influence each other. For example, an antioxidant may be much more effective at

Random Order Trial Number	Standard Order Trial Number	Response Observed Values y	Formulation F		Mixer Speed S		Temperature T	
			1 A	2 B	1 60	2 80	1 70	2 82
___	1	324	324		324		324	
___	2	334	334		334			334
___	3	324	324			324	324	
___	4	317	317			317		317
___	5	366		366	366		366	
___	6	371		371	371			371
___	7	360		360		360	360	
___	8	362		362		362		362
TOTAL		2758	1299	1459	1395	1363	1374	1384
NUMBER OF VALUES		16	8	8	8	8	8	8
AVERAGE		172.4	162.4	182.4	174.4	170.4	171.8	173.0
EFFECT			20.0		-4.0		1.2	

FIGURE 3.19 Response table for replication example

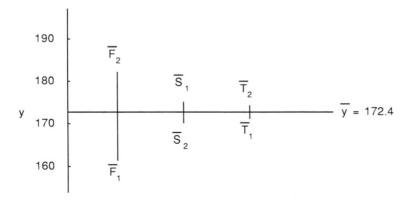

FIGURE 3.20 Graphical display of estimated effects

low temperatures than at high temperatures. When the effect of one factor is influenced by the level of another factor, we say there is an *interaction effect* between the two factors. In this section we will learn how to detect and measure such interaction effects among factors.

Orthogonal experimental designs protect against one factor causing an artificially large or small value for the estimated effect of another factor, but orthogonal designs do not always alert us to interactions between factors. That is, an estimated factor effect tells us the amount by which a response changes, *on the average,* when one factor goes from its low level to its high level. But suppose the average effect of factor A is different when factor B is at its low level than it is when factor B is at its high level. This interaction is averaged out when the average effect for factor A is calculated, and so the estimated factor A effect gives no clue about the interaction.

In section 3.3 the effect of a factor upon a response variable was estimated by calculating the average value of the response variable at the high level of the factor and also at the low value of the factor. The *effect* for the factor was the arithmetic difference between these two average values. Much the same thing is done to estimate interaction effects between factors. In order to find the interaction effect for factors A and B, say, the effect of factor A at

Factors A B	Average Response
1 1	25
1 2	31
2 1	36
2 2	42

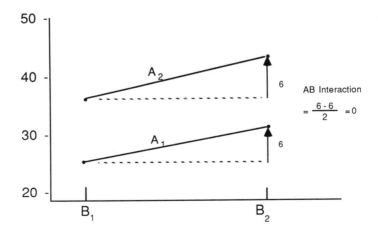

FIGURE 3.21 No interaction effect

Factors A B	Average Response
1 1	25
1 2	28
2 1	32
2 2	43

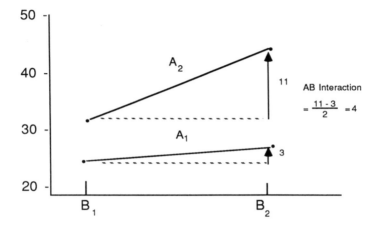

FIGURE 3.22 Positive AB interaction effect

the *high* level of B is calculated. Then the effect of A at the *low* level of B is calculated. Next the *difference* between these two effects is obtained. The interaction effect between A and B is, by convention, half the difference between these two effects.

This may seem a bit confusing, so let's look at a geometric interpretation of two-factor interaction. Figure 3.21 illustrates the case of no two-factor interaction. At level 1 for factor A, changing factor B from level 1 to level 2 causes the response to increase from 25 to 31, an increase of 6. At level 2 for factor A, changing B from level 1 to level 2 causes the response to increase from 36 to 42, again an increase of 6. So the effect of factor B is the same at both levels of factor A. That the two lines in Figure 3.21 are parallel immediately tells us the two factors do not interact.

Figure 3.22 shows two factors with a *positive* interaction effect: that is, B has a more positive effect (or less negative effect) at the high level of A than at the low level of A. We measure the interaction effect by finding the effect of B at level 1 of A $(28 - 25 = 3)$ and the effect of B at level 2 for A $(43 - 32 = 11)$. The *interaction effect* between A and B is half the difference of the B effects at levels 2 and 1 of factor A: $(11 - 3)/2 = 4$.

Figure 3.23 illustrates *negative* interaction. Here the effect of B at level 1

| Factors | | Average |
A	B	Response
1	1	25
1	2	28
2	1	43
2	2	32

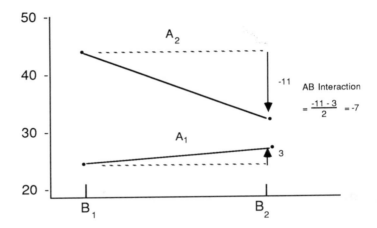

FIGURE 3.23 Negative AB interaction effect

of A is $28 - 25 = 3$ and the effect of B at level 2 of A is $32 - 43 = -11$. The interaction effect between A and B is then $(-11 - 3)/2 = -7$. We will encounter many other examples of interaction effects in this and later chapters. Interaction plots are discussed in several other texts, including Ott (1975), Daniel (1976), and Box, Hunter, and Hunter (1978).

It is not necessary to remember the difference between positive and negative interaction, or to remember the rationale for how interaction effects are calculated. We will use graphical techniques to identify and evaluate the practical implications of interactions. We will use a modified response table to simplify calculation of interaction effects. The response table is based on the algorithm described below.

Algorithm for calculating interaction effects

Calculation of interaction effects can be simplified using the following algorithm, presented by G.E.P. Box and J.S. Hunter (1961);

1. For a design matrix written in 1,2 notation, replace all 1s by −1s and 2s by 1s.
2. Multiply together the corresponding entries in the A and B effects columns. Put these products in a new column, labeled AB.
3. Repeat step 2 for the A and C columns and for the B and C columns.
4. Multiply together the values in the A column and the corresponding values in the y (response) column. Add these products and divide the sum by 4. This is the value for the A main effect. This step is equivalent to calculating the average responses at the high (+) and low (−) levels of A and obtaining the arithmetic difference between the average response at the high level of A minus the average difference at the low level of A.
5. Repeat step 4 for the B and C main effects.
6. Multiply together the values in the AB column and the corresponding values in the y column. Add these products and divide the sum by 4. This is the value for the AB interaction effect.
7. Repeat step 6 for the AC and BC interaction effects.

The columns of −1s and 1s for the A, B, and C main effects and AB, AC, and BC interaction effects are given in Figure 3.24. The reader might want to verify that the AB, AC, and BC are in fact obtained by multiplying together the corresponding elements in the A, B, and C columns. A response table based on Figure 3.24 is given in Figure 3.25 (the last column in Figure 3.25 will be explained shortly). Note that for each factor and each interaction there are two columns in Figure 3.25 in which to enter response values. The first column corresponds to −1s, and the second to 1s, in Figure 3.24. The calculations indicated by steps 4 through 7 above are accomplished (in slightly different order) by filling in a response table. The next example will show how this is done.

Example 3.3

Consider the data from a three-factor experiment given in Figure 3.26. The data have been entered into a response table in Figure 3.27. The grand aver-

Standard	MAIN EFFECTS			INTERACTION EFFECTS		
order	*A*	*B*	*C*	*AB*	*AC*	*BC*
1	−1	−1	−1	1	1	1
2	−1	−1	1	1	−1	−1
3	−1	1	−1	−1	1	−1
4	−1	1	1	−1	−1	1
5	1	−1	−1	−1	−1	1
6	1	−1	1	−1	1	−1
7	1	1	−1	1	−1	−1
8	1	1	1	1	1	1

FIGURE 3.24 Levels of interaction effects

Random Order Trial Number	Standard Order Trial Number	Response Observed Values y	A		B		C		AB		AC		BC		ABC	
			1	2	1	2	1	2	1	2	1	2	1	2	1	2
	1															
	2															
	3															
	4															
	5															
	6															
	7															
	8															
TOTAL																
NUMBER OF VALUES																
AVERAGE																
EFFECT																

FIGURE 3.25 Response table with interaction effects

	FACTOR			
Trial	A	B	C	y
1	1	1	1	24.7
2	1	1	2	21.4
3	1	2	1	44.1
4	1	2	2	43.5
5	2	1	1	38.4
6	2	1	2	32.0
7	2	2	1	40.2
8	2	2	2	37.5

FIGURE 3.26 Example design matrix and observed responses

age and three main effects have also been calculated in that figure. In Figure 3.28 the observed response values have been entered into the two-factor interaction columns and the interaction effects calculated. A graphical representation of the main and interaction effects appears in Figure 3.29. We see there that the dominant effects are B and the AB interaction. A and C are of less importance, and the AC and BC interactions appear negligible.

Suppose we wanted to maximize the value of the response for this ex-

Random Order Trial Number	Standard Order Trial Number	Response Observed Values y	A 1	A 2	B 1	B 2	C 1	C 2	AB 1	AB 2	AC 1	AC 2	BC 1	BC 2	1	2
6	1	24.7	24.7		24.7		24.7									
7	2	21.4	21.4		21.4			21.4								
2	3	44.1	44.1			44.1	44.1									
1	4	43.5	43.5			43.5		43.5								
8	5	38.4		38.4	38.4		38.4									
4	6	32.0		32.0	32.0			32.0								
3	7	40.2		40.2		40.2	40.2									
5	8	37.5		37.5		37.5		37.5								
TOTAL		281.8	133.7	148.1	116.5	165.3	147.4	134.4								
NUMBER OF VALUES		8	4	4	4	4	4	4								
AVERAGE		35.23	33.43	37.03	29.13	41.33	36.85	33.60								
EFFECT				3.60		12.20		-3.25								

FIGURE 3.27 Response table with data from Figure 3.26

Random Order Trial Number	Standard Order Trial Number	Response Observed Values y	A		B		C		AB		AC		BC			
			1	2	1	2	1	2	1	2	1	2	1	2	1	2
6	1	24.7	24.7		24.7		24.7			24.7		24.7		24.7		
7	2	21.4	21.4		21.4			21.4		21.4	21.4		21.4			
2	3	44.1	44.1			44.1	44.1		44.1			44.1	44.1			
1	4	43.5	43.5			43.5		43.5	43.5		43.5			43.5		
8	5	38.4		38.4	38.4		38.4		38.4		38.4			38.4		
4	6	32.0		32.0	32.0			32.0	32.0			32.0	32.0			
3	7	40.2		40.2		40.2	40.2			40.2	40.2		40.2			
5	8	37.5		37.5		37.5		37.5		37.5		37.5		37.5		
TOTAL		281.8	133.7	148.1	116.5	165.3	147.4	134.4	158.0	123.8	143.5	138.3	137.7	144.1		
NUMBER OF VALUES		8	4	4	4	4	4	4	4	4	4	4	4	4		
AVERAGE		35.23	33.43	37.03	29.13	41.33	36.85	33.60	39.50	30.95	35.88	34.58	34.43	36.03		
EFFECT			3.60		12.20		-3.25		-8.55		-1.30		1.60			

FIGURE 3.28 Response table with interaction terms

ample. Based on the estimates of main factor effects, and *ignoring for the moment any interaction effects*, we would decide that the average response would be maximized if we set factors A and B at their high levels and C at its low level. The overall mean for the eight observed values of the dependent

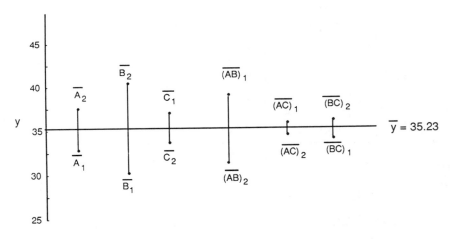

FIGURE 3.29 Graphical representation of main and interaction effects

LEVEL OF FACTORS		Observed values	Average
A	B		
1	1	24.7, 21.4	23.05
1	2	44.1, 43.5	43.80
2	1	38.4, 32.0	35.20
2	2	40.2, 37.5	38.85

FIGURE 3.30 Average responses by levels of A and B

variable is $\bar{y} = 35.23$, so the estimated maximum value for the response variable is:

$$\bar{y}_{max} = \text{(grand mean)} + \text{(A contribution)} + \text{(B contribution)} + \text{(C contribution)}$$
$$= 35.23 + (37.03 - 35.23) + (41.33 - 35.23) + (36.85 - 35.23)$$
$$= 44.75$$

But note in Figure 3.26 that for high A and B and low C, our observed value for y in the experiment was 40.2, while the maximum observed value for y was 44.1, which occurred when B was at its high level and A and C were at their low levels. This arouses suspicion. Actually, since the data showed a high AB interaction effect (see Figure 3.29), we should never have tried to maximize the response by looking only at the main effects of A, B, and C. We should, instead, calculate the average response at each combination of levels of A and B. This is done in Figure 3.30, and is based on the levels for A and B as given in Figure 3.26.

We can visually show the results of factor interaction by plotting the average values of the response variable at the four combinations of high and low values of two factors. In Figure 3.31 the AB, AC, and BC averages are plotted in three separate graphs. The values needed to plot Figure 3.31a are given in Figure 3.30. The values needed to plot Figures 3.31b and 3.31c can be obtained in a similar way. In Figure 3.31a a strong AB interaction is indicated. The two line segments being distinctly non-parallel indicates the presence of an interaction between the two factors. The nearly parallel lines in Figures 3.31b and 3.31c indicate that there is no strong interaction between A and C or between B and C.

Since A and B interact in this example, they should not be interpreted separately. Thus, if we want to determine what factor levels produce the largest average value for y, we should look at what combination of A and B produce the largest average y value. Since C did not appear to interact with either A or B, we can select the best value for C without regard to A or B. In Figure 3.30 the largest average value, 43.80, occurred when A was at its low value and B at its high value. The estimated maximum average value for y, based on the values for the three factors used in the experiment, is then

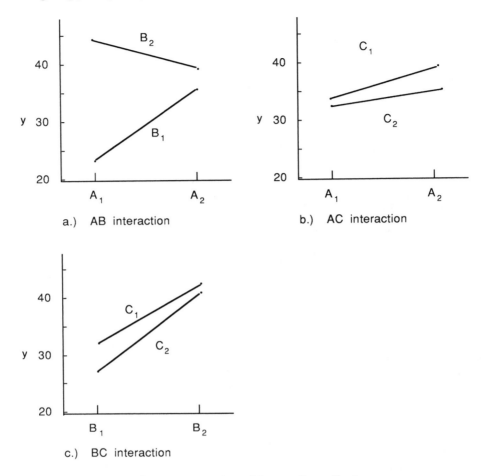

FIGURE 3.31 Graphical presentation of interaction effects

$$
\begin{aligned}
\bar{y}_{max} &= \text{(grand mean)} + \text{(A and B contribution)}^* \\
&\quad + \text{(C contribution)} \\
&= 35.23 + (43.80 - 35.23) + (36.85 - 35.23) \\
&= 35.23 + 8.57 + 1.62 \\
&= 45.42
\end{aligned}
$$

This is higher than the estimated maximum value of 44.75 which we obtained without considering the AB interaction. Any time there is evidence of interaction between factors, we should calculate the average value for y at each combination of these factors and work with these values.

---

*"(A and B contribution)" refers here to the A and B main effects *and* their interaction effect— here 43.80 is the average response when A is at level 1 and B is at level 2, so (43.80 − 35.23) is the "effect" on the average response of setting A and B at these two levels.

Three factor interactions

So far we have looked at three generic types of "effects":

- The grand mean: $\bar{y}$
- Main effects: A, B, and C,
- Two-way interaction effects: AB, AC, and BC.

One other type of interaction effect which is sometimes considered in two-level, three-factor experiments is the *three-way interaction* ABC. Calculating this effect is actually very easy. We add one more column to Figure 3.24 to represent ABC. This column is obtained by multiplying together, for each row, the values in the A, B, and C columns in that row. The resulting new column is the last column in Figure 3.32. This column is then used to "build" the last two columns in Figure 3.25. The three-factor interaction for the example we have been considering is calculated in Figure 3.33.

We usually calculate the estimated three-factor interaction effect at the same time we estimate the other effects. Interpreting the three-factor interaction term intuitively can be difficult. In section 3.6 we will use this term as part of the process for estimating basic variability of estimators.

3.6 Normal Plots of Estimated Effects

Suppose we toss a coin five times and get five straight heads. The probability of five straight heads in five tosses of a "fair" (perfectly balanced) coin is $2^{-5} = 1/32 = 0.03125 = 3.125$ percent. So if we were suspicious about the coin before we tossed it, we would be *very* suspicious after that little experiment. Suppose now that we had seven coins, and we tossed each one five times. If the coins are all fair, the chance of getting five straight heads or five straight tails for *at least* one of the coins can be shown mathematically to be $0.36 = 36\%$. So if for one of the coins we got all heads or all tails, we would not be surprised, and would not necessarily be suspicious of the coin involved. This little example epitomizes a problem which recurs throughout statistics: if you

Standard order	MAIN EFFECTS			INTERACTION EFFECTS			
	A	*B*	*C*	*AB*	*AC*	*BC*	*ABC*
1	-1	-1	-1	1	1	1	-1
2	-1	-1	1	1	-1	-1	1
3	-1	1	-1	-1	1	-1	1
4	-1	1	1	-1	-1	1	-1
5	1	-1	-1	-1	-1	1	1
6	1	-1	1	-1	1	-1	-1
7	1	1	-1	1	-1	-1	-1
8	1	1	1	1	1	1	1

FIGURE 3.32 Levels of interaction effects

Random Order Trial Number	Standard Order Trial Number	Response Observed Values y	A 1	A 2	B 1	B 2	C 1	C 2	AB 1	AB 2	AC 1	AC 2	BC 1	BC 2	ABC 1	ABC 2
6	1	24.7	24.7		24.7		24.7			24.7		24.7		24.7	24.7	
7	2	21.4	21.4		21.4			21.4		21.4	21.4		21.4			21.4
2	3	44.1	44.1			44.1	44.1		44.1			44.1	44.1			44.1
1	4	43.5	43.5			43.5		43.5	43.5		43.5			43.5	43.5	
8	5	38.4		38.4	38.4		38.4		38.4		38.4			38.4		38.4
4	6	32.0		32.0	32.0			32.0	32.0			32.0	32.0		32.0	
3	7	40.2		40.2		40.2	40.2			40.2	40.2		40.2		40.2	
5	8	37.5		37.5		37.5		37.5		37.5		37.5		37.5		37.5
TOTAL		281.8	133.7	148.1	116.5	165.3	147.4	134.4	158.0	123.8	143.5	138.3	137.7	144.1	140.4	141.4
NUMBER OF VALUES		8	4	4	4	4	4	4	4	4	4	4	4	4	4	4
AVERAGE		35.23	33.43	37.03	29.13	41.33	36.85	33.60	39.50	30.95	35.88	34.58	34.43	36.03	35.10	35.35
EFFECT			3.60		12.20		−3.25		−8.55		−1.30		1.60		0.25	

FIGURE 3.33 Response table with all interaction terms

do enough experimenting, "odd" results will eventually occur just by chance. In the illustrative example in section 3.5 it was found that the effects of B and AB were much larger than the others. Were these effects real, or just a chance result? Remember that we estimated seven effects (A, B, C, AB, AC, BC, and ABC). (We also tossed seven coins.) *Normal plots* is a graphical technique developed by Cuthbert Daniel (1959, 1976) to deal with this question. The technique is based on what is called the "Central Limit Theorem." This theorem says that, under fairly general conditions which we need not consider here, a weighted average of random measurements will be approximately normally distributed. In Chapter 5 the term "normally distributed" will be explained. Basically it means that a histogram of hundreds of weighted averages obtained under similar conditions would be "bell-shaped." Daniel built on this result by noting that if there were no "real" effects in an experiment, so that the differences among estimated effects were due only to chance, then we can reasonably expect certain patterns in the estimates. Moreover, these patterns can be examined graphically. The technique is very simple:

1. Order the estimated factor effects from smallest to largest. (Minus signs count here, so −8.5 is less than −1.6 or 2.8.)
2. Plot the points (E_i, P_i), i = 1, 2, . . . , m, on normal probability paper, where:

    m is the number of estimated effects,

    E_i is the *i*th smallest estimated factor effect,

    P_i is equal to $100\ (i - 0.5)/m$.

Figure 3.34 is a sheet of normal probability paper. Plot E_i on the horizontal axis and P_i on the vertical axis. For most of the two-level experiments considered in this book, m will be equal to either 7 or 15. Scales for P_i when m is 7 and 15 are given on the left side of Figure 3.34.

3. Fit a straight line through the set of points. Ignore points with very large or very small E_i values when fitting the line. Do any points with small E_i values appear to be too far to the left to "fit" the straight line? Do any

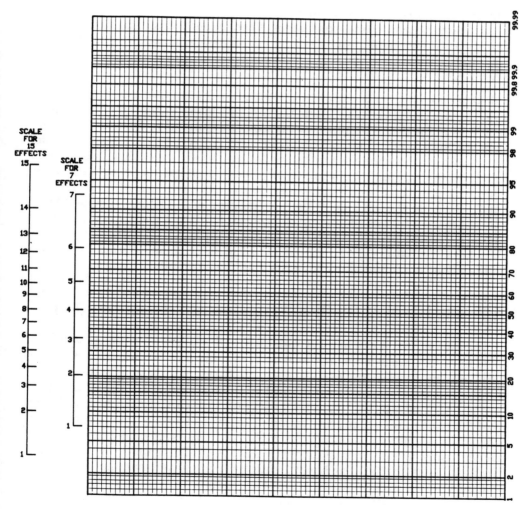

FIGURE 3.34 Normal probability paper

points with large E_i values appear to be too far to the right to fit the line? If the answer to either question is "yes", then these outlier points most likely represent real factor effects.

Points which are close to a line fitted to the middle group of points represent estimated factor effects which do not demonstrate any significant effect on the response variable. That is, if the point for a given estimated effect is close to the line plotted through the middle group of points, then we conclude that the present data do not give any indication that the factor has any significant effect on the behavior of the response variable. Of course, we must always make such statements cautiously. It is possible that data have so much variability that it is difficult to detect factor effects. Also, only two levels for each factor were included in the experiment. Perhaps a different selection of levels would give quite different results. In any event, we should never reach final conclusions based on one experiment. There should always be further experiments to confirm conclusions reached in any experiment.

Example

Consider the effects estimated in the illustrative example in section 3.5. The calculated values, E_i, of the factor effects are given at the bottom of Figure 3.33 and are also in column 2 of Figure 3.35. They have been ordered from smallest to largest in Figure 3.35. Their rank orders (order in size) are listed in column 3. In column 4 the values for $P_i = 100(i - 0.5)/7$ have been calculated. (Actually, since the number of estimated effects, m, is equal to 7, we didn't have to calculate the P_i values. We could have instead used the scale for 7 effects on the left side of the normal probability paper.) The points (E_i, P_i) for $i = 1, \ldots, 7$ are plotted on normal probability paper in Figure 3.36. A straight line has been fitted to the middle groupings of points. Note that the data indicate that the B and AB effects seem to be separate from the pattern of the other five points. This suggests that these two effects may be "real."

3.7 Mechanical Plating Experiment* (Example 3.4)

A manufacturer of mechanical fasteners had a chronic problem with zinc plating of lock washers being too thin. When this problem occurred, the usual "fix" was to increase the levels of cleaner, copper, and/or zinc in the "recipe" for the process. This action often resulted in increased plating costs while still producing inadequate plating thickness. An engineer at the plant suggested that the current recipe be reevaluated to see if the levels of one or more of the chemicals should be reduced. An eight-run, two-level experiment was designed using factor levels shown in Figure 3.37. To simplify the discussion, the new levels are listed as Level 1 since they are lower than the levels of the chemicals used in the existing recipe (Level 2).

During the experiment, only the above three controllable factors were allowed to vary. The following factors were held constant:

*We are indebted to Mr. Todd Schlei, Quality Systems Manager at Charter Manufacturing Company, for this example.

Factor	Estimated value, E_i	Rank order, i	$100(i - .5)/7$ P_i
AB	−8.55	1	7.1
C	−3.25	2	21.4
AC	−1.30	3	35.7
ABC	0.25	4	50.0
BC	1.60	5	64.3
A	3.60	6	78.6
B	12.20	7	92.3

FIGURE 3.35 Calculations for normal probability plot

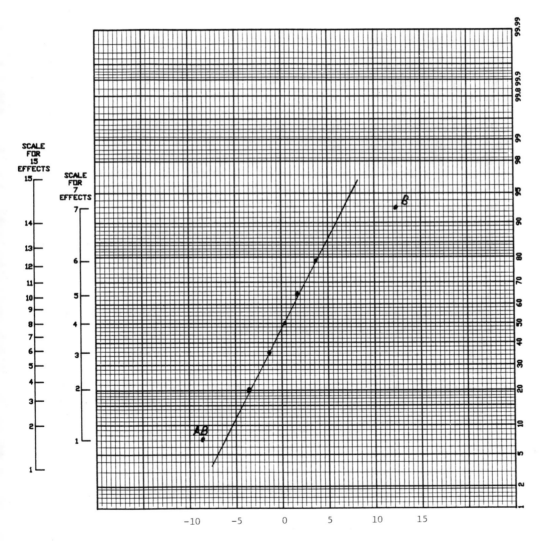

FIGURE 3.36 Normal plot of example effects

Factor	Level 1	Level 2
A: Cleaner	90% of recipe	recipe
B: Copper	71% of recipe	recipe
C: Zinc	67% of recipe	recipe

FIGURE 3.37 Levels of factors for experiment

a. same operator

b. same part number

c. same primary operations (that is, same machine and heat treatment furnace)

d. same timing for plating cycle

e. same load size (1,000 lbs.)

Thirty sample pieces were randomly selected from each experimental run. Thickness of coating was measured at three points on each sample piece. From these measurements an average thickness for the trial was calculated. These averages appear in the last column of Figure 3.38. Note that the order in which the trials were performed was randomized in Figure 3.38.

The completed response table for these data appears in Figure 3.39. The estimated effects are presented graphically in Figure 3.40. From this figure, it appears that thickness is maximized by setting factors A, B, and C at levels 1, 2, and 2 respectively: that is, decrease the level of cleaner, but hold the copper and zinc at their current recipe levels. We should be somewhat suspicious of this conclusion, however, because of the relatively large interaction effects. The effects are plotted on normal probability paper in Figure 3.41. There are no exceptional departures from the line fitted to the data, so at this point we begin to wonder if any of the factors we are studying have any effect on the response (thickness), at least over the ranges for these factors included in the experiment. If there are no factor effects, then we should recommend setting all factors at their lowest levels (Level 1), since these are the least expensive levels.

Figure 3.40, however, suggests that the interactions between A and B, and between A and C should perhaps be explored further. For comparison, we

Run order	Standard order	FACTORS			Plating thickness
		Cleaner	Copper	Zinc	
1	3	90%	100%	67%	0.00051
2	7	100%	100%	67%	0.00058
3	1	90%	71%	67%	0.00051
4	5	100%	71%	67%	0.00049
5	6	100%	71%	100%	0.00050
6	4	90%	100%	100%	0.00069
7	2	90%	71%	100%	0.00071
8	8	100%	100%	100%	0.00063

FIGURE 3.38 Completed report form for example

Random Order Trial Number	Standard Order Trial Number	Response Observed Values y	A Cleaner		B Copper		C Zinc		AB Cleaner and Copper		AC Cleaner and Zinc		BC Copper and Zinc		ABC	
			1	2	1	2	1	2	1	2	1	2	1	2	1	2
3	1	51	51		51		51			51		51		51	51	
7	2	71	71		71			71		71	71		71			71
1	3	51	51			51	51		51			51	51			51
6	4	69	69			69		69	69		69			69	69	
4	5	49		49	49		49		49		49			49		49
5	6	50		50	50			50	50			50	50		50	
2	7	58		58		58	58			58	58		58		58	
8	8	63		63		63		63		63		63		63		63
TOTAL		462	242	220	221	241	209	253	219	243	247	215	230	232	228	234
NUMBER OF VALUES		8	4	4	4	4	4	4	4	4	4	4	4	4	4	4
AVERAGE		57.8	60.5	55.0	55.3	60.3	52.3	63.3	54.8	60.8	61.8	53.8	57.5	58.0	57.0	58.5
EFFECT			-5.5		5.0		11.0		6.0		-8.0		0.5		1.5		

FIGURE 3.39 Response table for example (table values = actual values $\times 10^5$)

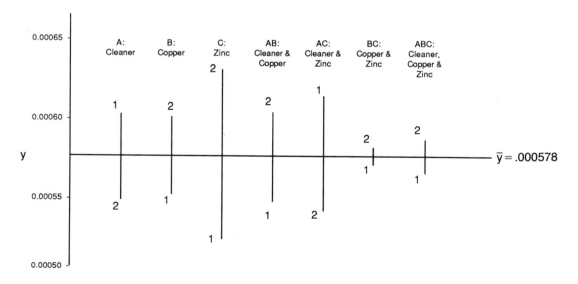

FIGURE 3.40 Graphical display of sample effects

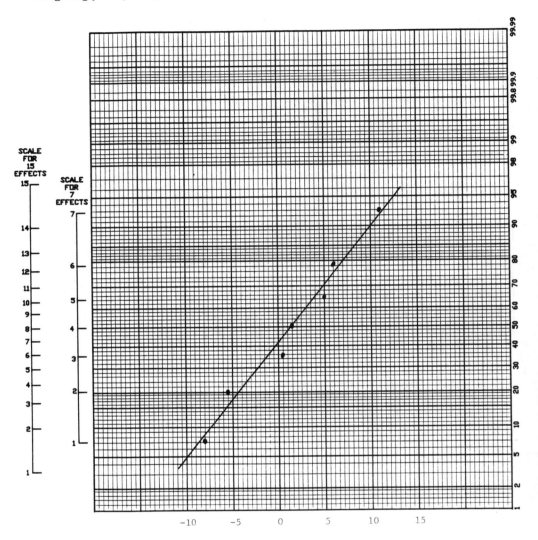

FIGURE 3.41 Normal plot of example effects

will also look at the interaction between B and C. The average responses for all combinations of two factors at a time are given in Figure 3.42. These averages are plotted in Figure 3.43 to show interaction effects. As expected (based on our earlier calculation of interaction effects), A and B, and A and C show interaction effects in Figure 3.43, while B and C do not. In order to maximize the response, Figure 3.43a says to set A at level 1. Factor B then has almost no effect on average response. Figure 3.43b tells us to set A at level 1 and C at level 2. Figure 3.43c also tells us to use level 2 for C, and to use level 2 for B as well. This is consistent with our earlier tentative decision before we considered interaction effects. The plant at which this experiment was performed decided to keep C (zinc) at level 2, but to switch both A (cleaner) *and*

	A: Cleaner	
	1	2
B: Copper 1	0.000610	0.000495
2	0.000600	0.000605

a.) Cleaner and Copper

	A: Cleaner	
	1	2
C: Zinc 1	0.000510	0.000535
2	0.000700	0.000565

b.) Cleaner and Zinc

	B: Copper	
	1	2
C: Zinc 1	0.000500	0.000545
2	0.000605	0.000660

c.) Copper and Zinc

FIGURE 3.42 Average responses at factor setting

B (copper) to level 1. The rationale for this decision on factor B was that when A was at level 1 and C at level 2, the level for B did not have a great effect on the response, and significant savings could be obtained by reducing the copper in the recipe (remember that Figure 3.41 told us not to take any of the estimated effects too seriously). Also, the largest observed response during the experiment occurred when A, B, and C were at levels 1, 1, and 2 respectively.

Verification runs were conducted at the recommended factor levels using different operators under normal production conditions. The observed thickness measurements are listed in Figure 3.44 and plotted in time order in Figure 3.45. Trials 5 and 6 had quite different results than the other six runs. Further investigation revealed that an automatic method of tumbling had been used for runs 5 and 6 while a manual tumbling procedure was used for the other six runs. Until this effect could be explored further, it was advised that automatic tumbling be discontinued.

A final report by the engineer who conducted this study concluded that, "Based on the results of the experiment, a substantial cost savings could be realized by using the optimum recipe. The experiment reversed some of the

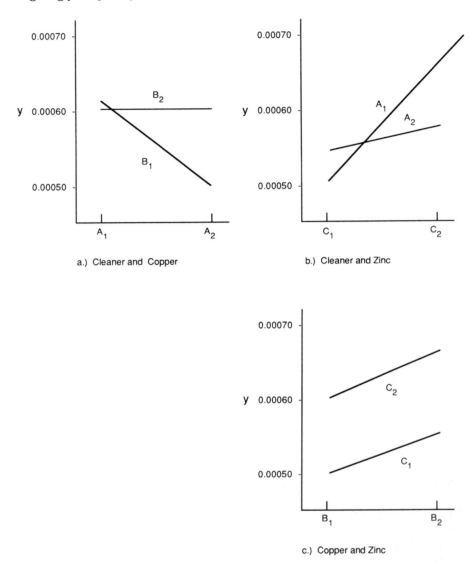

FIGURE 3.43 Graphs of interaction effects

thinking that originally controlled the process. Originally the thinking was that by using extra cleaner, the parts would accept the copper better, and by using extra copper, the zinc would adhere to the copper better thus giving you a greater thickness. The experiment has shown that there are certain points in the process that by adding additional chemicals, the result is less plating and a greater cost."

A critical concern which was not addressed in this example is the amount of *variability* in coating thickness. The experimenters focused exclusively on increasing *average* thickness without taking into consideration the effects the

Run number	Average thickness
1	0.000593
2	0.000656
3	0.000685
4	0.000700
5	0.000083
6	0.000088
7	0.000593
8	0.000578

FIGURE **3.44** Verification run data

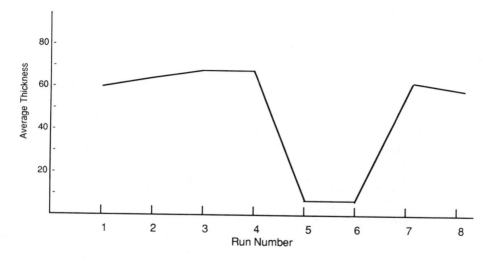

FIGURE 3.45 Time plot of verification runs for mechanical plating equipment

experimental factors might have on variability in thickness. Additional experiments should be performed (1) to explore the possibility of reducing cleaner and copper levels further, and (2) to determine the effects, if any, factor levels have on plating thickness variability. How to use experimental designs to assess effects on variability will be explained in Chapter 5.

3.8 Two-factor Design

The three-factor experimental design discussed in sections 3.3 through 3.7 is not an oddity or special case. Rather, it is one of a whole family of two-level experimental designs. We will look at two more "family members" in this chapter: a four-run experiment for two factors and a sixteen-run experiment for four factors. In the next chapter we will find that these three experimental

Standard order	FACTORS		Response (y)
	A	*B*	
1	1	1	y_1
2	1	2	y_2
3	2	1	y_3
4	2	2	y_4

FIGURE 3.46 Design matrix for two factors

designs will meet most of our needs for two-level designs for up to fifteen factors.

The design matrix for the four-run experiment in two factors is given in Figure 3.46. The response table form for this design is shown in Figure 3.47. This experiment is seldom run with just four trials. More often, it is replicated two or more times.

Example 3.5

A health-conscious engineer goes to a local fitness center three times each week to exercise. There are two different routes he can take from his home to the center. He wondered which route was faster, and also whether rush-hour traffic significantly affected travel time. He decided to resolve these issues by performing a four-run experiment, replicated three times. The design matrix

Random Order Trial Number	Standard Order Trial Number	Response Observed Values y	A		B		AB	
			1	2	1	2	1	2
	1							
	2							
	3							
	4							
TOTAL								
NUMBER OF VALUES								
AVERAGE								
EFFECT								

FIGURE 3.47 Response table for two-factor experiment in four runs

Random Order Trial Number	Standard Order Trial Number	Route	Rush Hour	Time
1	2	1	Y	14.7
2	3	2	N	9.5
3	1	1	N	10.6
4	4	2	Y	15.7
5	2	1	Y	15.2
6	1	1	N	11.5
7	4	2	Y	11.9
8	3	2	N	13.6
9	1	1	N	13.1
10	3	2	N	11.4
11	4	2	Y	13.8
12	2	1	Y	12.4

FIGURE 3.48 Completed report form for example

is as given in Figure 3.46 where factor A is route traveled (1 or 2) and factor B is whether travel is during rush hour (1 = No, 2 = Yes). The completed report form for the experiment is in Figure 3.48. The response data are partially summarized in Figure 3.49. These data have been entered into the response table in Figure 3.50. Note that the "Number of Values" measurements in Figure 3.50 reflect the replications. The estimated effects are plotted in Figure 3.51. Based on this figure we conclude that:

Standard Order Trial Number	Route	Rush Hour	Responses R_1	R_2	R_3	Sum
1	1	N	10.6	11.5	13.1	35.2
2	1	Y	14.7	15.2	12.4	42.3
3	2	N	9.5	13.6	11.4	34.5
4	2	Y	15.7	11.9	13.8	41.4

FIGURE 3.49 Summary of report form data

Random Order Trial Number	Standard Order Trial Number	Response Observed Values y	A ROUTE		B RUSH HOUR		AB	
			1	2	No	Yes		
			1	2	1	2	1	2
	1	35.2	35.2		35.2			35.2
	2	42.3	42.3			42.3	42.3	
	3	34.5		34.5	34.5		34.5	
	4	41.4		41.4		41.4		41.4
TOTAL		153.4	77.5	75.9	69.7	83.7	76.8	76.6
NUMBER OF VALUES		12	6	6	6	6	6	6
AVERAGE		12.78	12.92	12.65	11.62	13.95	12.80	12.77
EFFECT			−0.27		2.33		−0.03	

FIGURE 3.50 Response table for example

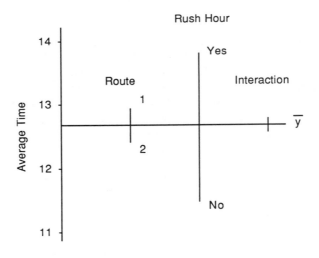

FIGURE 3.51 Plot of example estimated effects

- Route has a negligible effect on travel time;
- Travel time during rush hour is about 2.33 minutes longer, on the average, than travel time during off-hours;
- There is no significant interaction between route and rush hour. That is, rush hour has the same effect on each route.

3.9 Four-factor Design

With a sixteen-run experiment it is possible to estimate the main effects and all interaction effects for four factors. Additionally, all the estimates are orthogonal. Although this may sound ideal, it is actually overkill. In industrial experiments we are seldom interested in three-factor interactions, much less four-factor interactions; the ability to estimate the ABC, ABD, ACD, BCD, and ABCD interactions seems an unnecessary luxury. However, being able to include all these estimates in a normal plot is a distinct advantage when attempting to sort out the "real" effects from random variation. In Chapter 4 we will find some other uses for this design as well. The design matrix for a four-factor, two-level, sixteen-run experiment is given in Figure 3.52 using the 1 = low and 2 = high notation. The factor levels and corresponding levels for interaction terms are given in Figure 3.53 using the -1 = low and 1 = high notation. In section 3.5 we generated the levels for interaction effects for a three-factor experiment by multiplying together the values in corresponding main effects columns of the design matrix. This can be done whenever the -1 and 1 notation is used. For example, in Figure 3.53 the entries in the "AB" column are equal to the product of the terms in the A and B columns. Similarly, the values in the ABC column are obtained by multi-

	Levels of Factors				
Standard Order	A	B	C	D	Response
1	1	1	1	1	y_1
2	1	1	1	2	y_2
3	1	1	2	1	y_3
4	1	1	2	2	y_4
5	1	2	1	1	y_5
6	1	2	1	2	y_6
7	1	2	2	1	y_7
8	1	2	2	2	y_8
9	2	1	1	1	y_9
10	2	1	1	2	y_{10}
11	2	1	2	1	y_{11}
12	2	1	2	2	y_{12}
13	2	2	1	1	y_{13}
14	2	2	1	2	y_{14}
15	2	2	2	1	y_{15}
16	2	2	2	2	y_{16}

FIGURE 3.52 Design matrix for sixteen-run experiment

Standard Order	Main Factors				Interactions										
	A	B	C	D	AB	AC	AD	BC	BD	CD	ABC	ABD	ACD	BCD	ABCD
1	-1	-1	-1	-1	1	1	1	1	1	1	-1	-1	-1	-1	1
2	-1	-1	-1	1	1	1	-1	1	-1	-1	-1	1	1	1	-1
3	-1	-1	1	-1	1	-1	1	-1	1	-1	1	-1	1	1	-1
4	-1	-1	1	1	1	-1	-1	-1	-1	1	1	1	-1	-1	1
5	-1	1	-1	-1	-1	1	1	-1	-1	1	1	1	-1	1	-1
6	-1	1	-1	1	-1	1	-1	-1	1	-1	1	-1	1	-1	1
7	-1	1	1	-1	-1	-1	1	1	-1	-1	-1	1	1	-1	1
8	-1	1	1	1	-1	-1	-1	1	1	1	-1	-1	-1	1	-1
9	1	-1	-1	-1	-1	-1	-1	1	1	1	1	1	1	-1	-1
10	1	-1	-1	1	-1	-1	1	1	-1	-1	1	-1	-1	1	1
11	1	-1	1	-1	-1	1	-1	-1	1	-1	-1	1	-1	1	1
12	1	-1	1	1	-1	1	1	-1	-1	1	-1	-1	1	-1	-1
13	1	1	-1	-1	1	-1	-1	-1	-1	1	-1	-1	1	1	1
14	1	1	-1	1	1	-1	1	-1	1	-1	-1	1	-1	-1	-1
15	1	1	1	-1	1	1	-1	1	-1	-1	1	-1	-1	-1	-1
16	1	1	1	1	1	1	1	1	1	1	1	1	1	1	1

FIGURE 3.53 Levels for factors and interactions

plying together the corresponding terms in the A, B, and C columns. At this point we do not need to use Figure 3.53, but it will be important in Chapter 4. A response table form for this design is given in Figure 3.54.

Example 3.6

A milling operation was evaluated for finish milling of a particular part. The critical controllable factors affecting surface finish were identified as cutting speed, depth of cut, cutter diameter, and feed per tooth. Effects of these factors were estimated using a sixteen-run experiment with two levels for each factor. The factor levels used are given in Figure 3.55. A completed report form for the experiment is given in Figure 3.56. The data are recorded in the random order in which the runs were performed. For this experiment, smaller values are better (smoother). A completed response table for the experiment is given in Figure 3.57.

The effects calculated at the bottom of Figure 3.57 are displayed graphically in Figure 3.58. The key effects, in order of impact on smoothness, are D, BD, AD, A, and B. C does not seem to have any real effect on smoothness. Ignoring for the moment the interaction effects, the experiment indicates that the best settings for factors A, B and D are:

$$\text{Cutter speed} = A_2 = 100 \text{ m/min,}$$
$$\text{Depth of cut} = B_1 = 1 \text{ mm,}$$
$$\text{Feed/tooth} = D_1 = 0.25 \text{ mm/tooth.}$$

Random Order Trial Number | Standard Order Trial Number | Response Observed Values y

| 1 | 2 | 1 | 2 | 1 | 2 | 1 | 2 | 1 | 2 | 1 | 2 | 1 | 2 | 1 | 2 | 1 | 2 | 1 | 2 | 1 | 2 | 1 | 2 | 1 | 2 | 1 | 2 | 1 | 2 | 1 | 2 |

1
2
3
4
5
6
7
8
9
10
11
12
13
14
15
16

TOTAL

NUMBER OF VALUES

AVERAGE

EFFECT

FIGURE 3.54 Response table for sixteen-run experiment

Level	Factors			
	Cutting speed (m/min)	Depth of cut (mm)	Cutter diameter (mm)	Feed/tooth (mm/tooth)
High	100	8	200	0.65
Low	80	1	100	0.25

FIGURE 3.55 Factor levels for sixteen-run experiment

Figure 3.59 is a normal plot of the fifteen effects estimated in Figure 3.57. In this plot only the five effects noted above (D, BD, AD, A, and B) deviate significantly from the fitted straight line. Since the AD and BD interaction effects appear to be important to the response variable (surface finish) the average values of the response should be calculated and plotted for each combination of levels of A and D, and also for each combination of levels of B and D. The average responses are calculated in Figures 3.60 and 3.62. The average responses are plotted in Figures 3.61 and 3.63.

Figure 3.61 indicates that in order to minimize average response, both D

Run order for experiment	Standard order of trials	A: Cutting speed (m/min)	B: Depth of cut (mm)	C: Cutter diameter (mm)	D: Feed/ tooth (mm/tooth)	Y: Surface finish (microinches)
1	11	100	1	200	0.25	49
2	5	80	8	100	0.25	74
3	13	100	8	100	0.25	67
4	2	80	1	100	0.65	101
5	10	100	1	100	0.65	70
6	16	100	8	200	0.65	65
7	9	100	1	100	0.25	44
8	12	100	1	200	0.65	65
9	8	80	8	200	0.65	95
10	7	80	8	200	0.25	69
11	3	80	1	200	0.25	46
12	14	100	8	100	0.65	60
13	6	80	8	100	0.65	86
14	15	100	8	200	0.25	80
15	1	80	1	100	0.25	42
16	4	80	1	200	0.65	103

FIGURE 3.56 Example report form

Response table for a 2⁴ factorial design (example data).

Random Order Trial Number	Standard Order Trial Number	Response Observed Values y	A: Cutting Speed 80/100 (1)	A (2) 100	B: Depth of Cut 1 (1)	B (2) 8	C: Cutter Diameter 100/200 (1)	C (2)	D: Feed/Tooth .25 (1)	D (2) .65	AB (1)	AB (2)	AC (1)	AC (2)	AD (1)	AD (2)	BC (1)	BC (2)	BD (1)	BD (2)	CD (1)	CD (2)	ABC (1)	ABC (2)	ABD (1)	ABD (2)	ACD (1)	ACD (2)	BCD (1)	BCD (2)	ABCD (1)	ABCD (2)
15	1	42	42		42		42		42		42		42		42		42		42		42		42		42		42		42		42	
4	2	101	101		101		101			101		101	101			101		101	101		101			101	101			101		101	101	
11	3	46	46		46		46		46		46			46	46		46		46		46		46		46		46			46		46
16	4	103	103		103		103			103	103		103		103		103			103		103		103		103	103		103			103
2	5	74	74			74	74		74			74	74			74	74		74		74			74	74			74	74			74
13	6	86	86			86	86		86			86		86	86			86	86		86		86	86			86	86		86		86
10	7	69	69			69	69		69		69			69		69	69			69	69		69	69		69		69		69	69	
9	8	95	95			95		95	95		95			95	95		95			95		95		95	95		95			95	95	
7	9	44		44	44		44		44		44		44		44			44	44			44	44			44		44	44			44
5	10	70		70	70		70		70		70			70	70			70	70			70	70			70	70			70	70	
1	11	49		49	49		49		49		49		49			49	49		49		49			49	49			49	49			49
8	12	65		65	65		65		65			65	65			65	65		65			65	65		65		65		65		65	
3	13	67		67	67			67	67			67	67		67			67	67		67		67		67			67	67			67
12	14	60		60	60		60		60		60		60		60		60		60		60		60		60		60			60	60	
14	15	80		80	80			80	80		80			80		80	80			80	80		80			80	80			80		80
6	16	65		65	65		65		65		65		65		65		65		65		65		65		65		65		65		65	
TOTAL		1116	616	500	520	596	544	572	471	645	552	564	554	562	625	491	550	566	629	487	561	555	548	568	551	565	573	543	549	567	565	551
NUMBER OF VALUES		16	8	8	8	8	8	8	8	8	8	8	8	8	8	8	8	8	8	8	8	8	8	8	8	8	8	8	8	8	8	8
AVERAGE		69.8	77.0	62.5	65.0	74.5	68.0	71.5	58.9	80.6	69.0	70.5	69.3	70.3	78.1	61.4	68.8	70.8	78.6	60.9	70.1	69.4	68.5	71.0	68.9	70.6	71.6	67.9	68.6	70.9	70.6	68.9
EFFECT			-14.5		9.5		3.5		21.7		1.5		1.0		-16.7		2.0		-17.7		-0.7		2.5		1.7		-3.7		2.3		-1.7	

FIGURE 3.57 Response table for example data

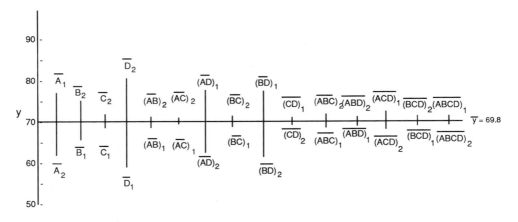

FIGURE 3.58 Graphical display of example effects

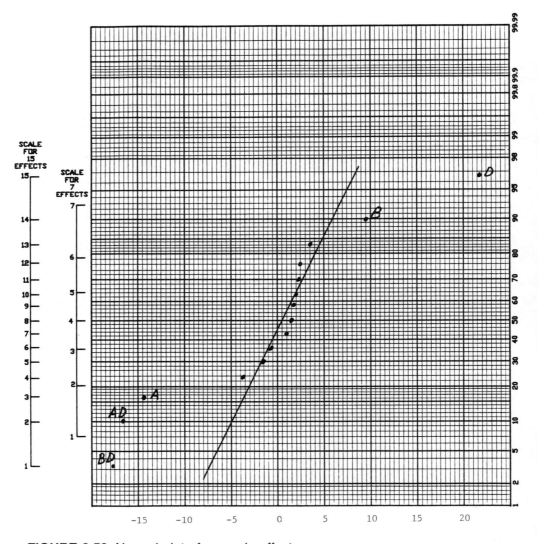

FIGURE 3.59 Normal plot of example effects

		D: Feed per tooth	
		1	**2**
A: Cutting Speed	**1**	42 46 74 69 total = 231 $\overline{A_1 D_1}$ = 231/4 = 57.8	101 103 86 95 total = 385 $\overline{A_1 D_2}$ = 385/4 = 96.3
	2	44 49 67 80 total = 240 $\overline{A_2 D_1}$ = 240/4 = 60.0	70 65 60 65 total = 260 $\overline{A_2 D_2}$ = 260/4 = 65.0

FIGURE 3.60 Average response as a function of levels of factors A and D

and A should be set at their low levels. (Recall that when we considered just average responses based on factor levels without regard to interactions, the data indicated that the low level of D *and high level of* A should be used. But we need to be careful here. Note in Figure 3.61 that when A is at its high level, A_2, the average response is not as sensitive to change in level of D as it is when A is at its low level (A_1). If the level of D is hard to control, or if D

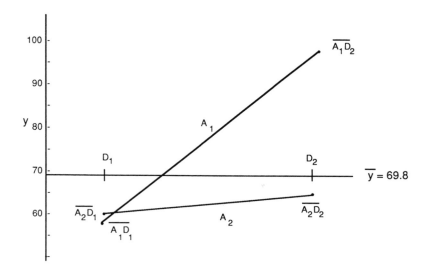

FIGURE 3.61 Plot of average response as function of levels of factors A and D

		D: Feed per tooth	
		1	**2**
B: Depth of Cut	**1**	42 46 44 49 total = 181 $\overline{B_1 D_1}$ = 181/4 = 45.3	101 103 70 65 total = 339 $\overline{B_1 D_2}$= 339/4 = 84.8
	2	74 69 67 80 total = 290 $\overline{B_2 D_1}$= 290/4 = 72.5	86 95 60 65 total = 306 $\overline{B_2 D_2}$= 306/4 = 76.5

FIGURE 3.62 Average response as a function of levels of factors B and D

is a noise factor, then it would probably be better to set A at its high level as a way of reducing process variability. The "Taguchi approach" to quality engineering places heavy emphasis on using experiments during parameter design to make products and processes robust to noise factors. Figure 3.63 is another interesting situation. Variability of the average response, as a func-

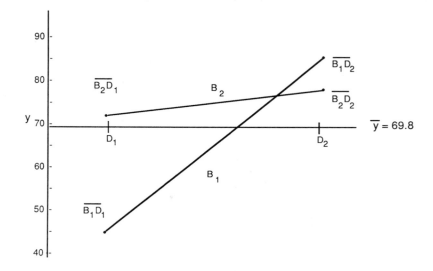

FIGURE 3.63 Plot of average response as a function of levels B and D

tion of D, is lower when B is at its high level, but the average response is much lower when B and D are both at their lower values. It appears that A, B, and D should be set at levels 2, 1, and 1, respectively, but that further study may be needed of the interaction of A and B with D.

Levels 2, 1, and 1 of factors A, B, and D correspond to settings of 100 m/min, 1 mm, and 0.25 mm/tooth for the three factors (see Figure 3.55). Those factor levels were used in trials (row numbers) 1 and 7 in Figure 3.56. The observed responses (y) for those two trials were 49 and 44. The observed average response at levels 2, 1, and 1 for factors A, B, and D is then $(49+44)/2 = 46.5$. This is $69.8 - 46.5 = 23.3$ microinches less than the observed overall response.

3.10 Taguchi Designs and Western Designs

There has been considerable debate about the differences between "traditional" or "Western" experimental design and Taguchi's approach to off-line quality engineering. Unfortunately, a few people have blown these differences out of proportion and claim that either Western designs are ineffective or Taguchi designs are invalid. The truth is that the designs Taguchi uses have been used by statisticians, engineers and scientists for years. But Taguchi made these designs more appealing to engineers by presenting them with case studies rather than mathematical formulas. Dr. Taguchi avoided the tendency common among Western statisticians of pointing out the technical merits and theoretical elegance of certain systems of experimental designs, at the cost of clarity of how to use the tools.

Figure 3.64 includes two design matrices. The one on the left is usually called a 2^3 design by Western statisticians. On the right is Taguchi's L_8 design. These two designs are really the same. The only differences have to do with notation and ordering. They are also both the same as the eight-run design we studied in this chapter. Here are the differences/similarities:

- All three designs have eight runs and seven orthogonal columns in their factor matrices.

WESTERN 2^3 DESIGN							TAGUCHI L_8 DESIGN						
A	*B*	*C*	*AB*	*AC*	*BC*	*ABC*	*1*	*2*	*3*	*4*	*5*	*6*	*7*
−1	−1	−1	1	1	1	−1	1	1	1	1	1	1	1
1	−1	−1	−1	−1	1	1	1	1	1	2	2	2	2
−1	1	−1	−1	1	−1	1	1	2	2	1	1	2	2
1	1	−1	1	−1	−1	−1	1	2	2	2	2	1	1
−1	−1	1	1	−1	−1	1	2	1	2	1	2	1	2
1	−1	1	−1	1	−1	−1	2	1	2	2	1	2	1
−1	1	1	−1	−1	1	−1	2	2	1	1	2	2	1
1	1	1	1	1	1	1	2	2	1	2	1	1	2

FIGURE 3.64 Comparison of Taguchi and Western two-level, eight-run designs

- All three designs have columns for each of their three main effects, three two-factor interactions and one three-factor interaction.
- If, for the eight-run experimental design we have been considering in this chapter, we interchange columns 1 and 3, and columns 4 and 6, we have the Western 2^3 design. To verify that this is true, compare the matrix on the left in Figure 3.64 with Figure 3.32. What is involved here is just interchanging factors A and C.
- Consider the eight-run experiment of this chapter, as expressed in Figure 3.32. Replace the -1s by 2s, leave the 1s alone, interchange columns 3 and 4, and reverse the order of the runs. This gives us the Taguchi L_8 experiment as it appears on the right in Figure 3.64.

So the differences in these experimental designs are simply a matter of notation, not substance.

The similarities/differences with the eight-run designs have parallels in the four-run and sixteen-run designs of section 3.8. Consider the three two-factor matrices in Figure 3.65. If, for the Western design in Figure 3.65, the -1s are replaced by 1s, and the 1s replaced by 2s, and the first two columns interchanged, we obtain the left-most matrix in Figure 3.65. If, for the Taguchi design in Figure 3.65, we interchange the 1s and 2s, and reverse the order of the rows, we again obtain the left-most matrix in Figure 3.65. So once again, we have differences in notation, not substance.

For sixteen-run experiments the same sort of situation applies. In particular, we have the assignment of columns to effects expressed in Figure 3.66. Figure 3.53 is just the Western 2^4 design with columns rearranged as indicated in Figure 3.66. To get the Taguchi L_{16} design from Figure 3.53 we would have to replace -1s by 2s and reverse the ordering of the rows in Figure 3.53, in addition to making the column switches indicated in Figure 3.66.

If the designs are really the same, why didn't we pick either the Western notation or the Taguchi notation and stick with it? At first we were going to use the Taguchi notation, because of its recent popularity. But this would make it more difficult for readers to go on after reading this book and learn more about experimental designs. Taguchi uses only a few of the experimental designs developed by Western statisticians. In time many readers will want to learn about some of the other designs. We also considered using the Western notation exclusively, but decided that many readers would already have had some introduction to Taguchi methods, and we wanted to make learning

SECTION 3.8			WESTERN 2^3			TAGUCHI L_8		
A	*B*	*AB*	*A*	*B*	*AB*	*1*	*2*	*3*
1	1	2	-1	-1	1	1	1	1
1	2	1	1	-1	-1	1	2	2
2	1	1	-1	1	-1	2	1	2
2	2	2	1	1	1	2	2	1

FIGURE 3.65 Forms of a four-run, two-level design

Effect	Figure 3.53	Taguchi L_{16}	Western 2^4
A	1	1	4
B	2	2	3
C	3	4	2
D	4	8	1
AB	5	3	10
AC	6	5	9
AD	7	9	7
BC	8	6	8
BD	9	10	6
CD	10	12	5
ABC	11	7	14
ABD	12	11	13
ACD	13	13	12
BCD	14	14	11
ABCD	15	15	15

FIGURE 3.66 Assignment of effects to columns of design matrices

as easy for this group as possible. Hence the format we have selected. The response tables we have introduced can be comfortably used by engineers with background in either style of notation. These forms intentionally avoid drawing attention to the notation being used for the design, and allow us to use currently available Western software packages.

4

Two-level Experiments: Fractional Factorial Designs

4.1 Fractional Factorial Designs Based on Eight-run Experiments

In Chapter 3 we saw how to analyze the effects of two factors in $2^2 = 4$ trials; how to analyze the effects of three factors in $2^3 = 8$ trials; and how to analyze the effects of four factors in $2^4 = 16$ trials. But must we always use 2^k trials when working with k factors? The answer, fortunately, is no. In fact, we can sometimes analyze the effects of k factors in only $k+1$ trials. But we must pay a price for this—when fewer than 2^k trials are used to analyze the effects of k factors, not all interaction effects can be cleanly estimated, and main factors may be "confounded" (a term which will be explained shortly) with interaction terms. However, through careful advanced planning experiments can be designed which require minimal numbers of trials and still provide estimates of the most important interaction effects.

Recall the eight-run experimental design presented in section 3.3. The "design matrix" for this design, which shows the levels of the three factors for each trial in the design, was given in Figure 3.6. The levels for the three factors are also given, using the $-1 =$ low and $1 =$ high notation, in columns 2, 3, and 4 of Figure 3.32 and are here reproduced in Figure 4.1. Figures 3.32 and 4.1 also show the levels of the interaction effects used in the experiment. The $-1, 1$ notation will be used extensively in this section since it facilitates our understanding of the material.

Four factors in eight trials

Suppose we want to evaluate the effects of four factors, at two levels each, using only eight trials. One way to do this would be to relabel the ABC inter-

Standard	MAIN EFFECTS			INTERACTION EFFECTS			
order	*A*	*B*	*C*	*AB*	*AC*	*BC*	*ABC*
1	−1	−1	−1	1	1	1	−1
2	−1	−1	1	1	−1	−1	1
3	−1	1	−1	−1	1	−1	1
4	−1	1	1	−1	−1	1	−1
5	1	−1	−1	−1	−1	1	1
6	1	−1	1	−1	1	−1	−1
7	1	1	−1	1	−1	−1	−1
8	1	1	1	1	1	1	1

FIGURE 4.1 Factor and interaction levels for a three-factor experiment in eight trials

action column as being the fourth factor column. That is, factors A, B, and C would have their levels determined by the values in the columns assigned to them in Figure 4.1. The fourth factor, D, would be set at its low or high level in each trial according to the value in the D = ABC column. This gives the design matrix in Figure 4.2. The trial values (−1 or 1) for the interaction columns can then be obtained by multiplying together corresponding column values, as was done in section 3.5. The columns for the interaction terms involving A, B, and C without D are given in Figure 4.1. The interaction terms involving D are AD, BD, CD, ABD, ACD, BCD, and ABCD. The columns for the two-factor interaction effects involving D are listed in the last three columns of Figure 4.3.

If we compare the columns in Figures 4.1 and 4.3, we find that not only is ABC = D, but also AB = CD. That is, the sequence of 1s and −1s in the AB column of Figure 4.1 is identical to the sequence of 1s and −1s in the CD column in Figure 4.3. So when estimating the CD effect, the same sequence of observed responses is added and subtracted as when estimating the AB effect. The two estimated effects will be equal. Similarly, AC = BD and BC = AD. Note that these "equations" can be obtained by switching letters in the equation ABC = D. Based on this discovery, we might wonder if perhaps A = BCD, B = ACD and C = ABD as well. In fact these relationships do hold. The reader

A	*B*	*C*	*D*
−1	−1	−1	−1
−1	−1	1	1
−1	1	−1	1
−1	1	1	−1
1	−1	−1	1
1	−1	1	−1
1	1	−1	−1
1	1	1	1

FIGURE 4.2 Design matrix for four factors in eight trials

A	B	C	D	CD	BD	AD
−1	−1	−1	−1	1	1	1
−1	−1	1	1	1	−1	−1
−1	1	−1	1	−1	1	−1
−1	1	1	−1	−1	−1	1
1	−1	−1	1	−1	−1	1
1	−1	1	−1	−1	1	−1
1	1	−1	−1	1	−1	−1
1	1	1	1	1	1	1

FIGURE 4.3 Some of the factor levels for the experiment defined by Figure 4.2

may want to verify some of these using values from columns in Figures 4.1 and 4.3. The reader might also want to discover the column values for the four-factor interaction term ABCD by muliplying together the AB and CD columns, or the A, B, C, and D columns, or any other combination of A, B, C, and D. Our ability to move letters around in the manner described above is based on sound theory, which will not be discussed in this book. For further reading on this topic see Box and Hunter (1961) and Box, Hunter, and Hunter (1978).

What are the practical implications of setting D equal to ABC? The most serious result is *confounding* of effects, which means being unable to determine which of two or more effects may be affecting the response variable. D is confounded with ABC, AB is confounded with CD, etc. When two effects are confounded, we say that each is an *alias* of the other. As a simple example of confounding, suppose we want to know which of two assembly methods was faster. We have Ann perform the assembly using method A and Dan perform the assembly using method B. The study shows that Ann is able to assemble an average of 167 units per hour, while Dan can only assemble 153 units per hour. Is method A faster than method B, or is Ann a faster assembler than Dan? We don't know, since assembly method and worker are confounded in this experiment. If Ann and Dan had each used method A half the time and method B half the time, the confounding would have been avoided. Suppose parts from supplier A do not rust as quickly as parts from supplier B. We find that supplier A uses a protective paint of type X and supplier B uses type Y protective paint. Is type X better than type Y? We don't know. Performance can be affected by surface preparation, method of application, method of drying, and a whole host of other factors which are confounded with type of coating. Again, only a properly designed and controlled experiment can provide the answer.

As we just saw, the alias structure created by assigning D to the ABC column in an eight-run experiment means that:

D is confounded with ABC

C is confounded with ABD

B is confounded with ACD

A is confounded with BCD

AB is confounded with CD

AC is confounded with BD

AD is confounded with BC

Thus, if the estimated effect for AB is large, we may not know if it is the AB interaction or the CD interaction which is affecting the response variable. However, experience has shown that real interaction effects are not likely to occur unless at least one of the factors involved in the interaction separately has an effect. Also, only rarely will significant three- and four-factor interactions be observed. Some experimenters ignore all interactions involving more than two factors, and assume that real two-factor interactions will not occur unless both factors involved in the interaction have significant main effects. This may seem a bit risky, but errors in decisions based on these rules often surface later during confirmatory experiments. But no criteria like these should be followed blindly. The first consideration in sorting out which interactions can or cannot occur is the existing knowledge about the experimental factors, particularly as they interact or do not interact with each other.

When confronted with confounding, determination of which factors are really affecting the response variable can sometimes be handled by dealing with the information as a whole. For example, suppose the experiment described in Figure 4.2 is performed, and the estimated factor effects listed in Figure 4.4 obtained. What effects are real? The data show large B = ACD, D = ABC and AC = BD effects. Since we do not expect to see three-factor interactions very often, we would decide that factors B and D are affecting the response, rather than the ACD and ABC interaction effects. This conclusion is supported by the low estimates for the A and C main effects. In the absence of other engineering knowledge about the relationships among the factors, the large AC = BD effect in Figure 4.4 should be attributed to a BD interaction rather than an AC interaction since B and D have large estimated main effects, while A and C do not. But what if we "guess wrong?" There are no guarantees in engineering experiments. But the chances of reaching incorrect decisions can be minimized by properly using appropriate experimental designs,

Factor	Alias	Estimate
A	BCD	4.8
B	ACD	18.2
C	ABD	2.9
D	ABC	23.4
AB	CD	3.0
AC	BD	11.5
AD	BC	5.1

FIGURE 4.4 Illustrative estimated effects and alias relationships for an eight-run experiment

carefully interpreting information on confounding of effects, and *always* following up experiments with confirmatory tests.

Example 4.1

An engineer wanted to maximize the bond strength when mounting an integrated circuit (I.C.) on a metalized glass substrate. Four factors are identified as potentially affecting the strength: adhesive type, conductor material, cure time, and I.C. post coating. The engineer decided to carry out an eight-run experiment using two levels for each of the four factors. The response variable would be bond strength, measured in pounds. Figure 4.5 gives the levels selected for the experimental factors. The factor levels used for the factors should be as given in Figure 4.2, with "-1" representing low levels and "1" representing high levels. The completed report form, but with trials listed in standard order instead of in the order the trials were performed, is given in Figure 4.6. Note the levels for the factors in Figure 4.6 are equivalent to the levels indicated by Figure 4.2. Figure 4.7 is a response table for the experiment. Factor D goes in the last columns of the table since we set $D = ABC$. The AB, AC, and BC columns could also be given the headings of CD, BD, and AD, respectively, due to the alias structure for this experiment, discussed earlier in this section.

The effects calculated in Figure 4.7 are plotted in Figure 4.8. This graph indicates that factors C (cure time) and D (I.C. post coating), and possibly

Factor	Low level	High level
A. Adhesive type	D2A	H-1-E
B. Conductor material	copper	nickel
C. Cure time (at 90° C)	90 min	120 min
D. IC post coating	tin	silver

FIGURE 4.5 Factor levels for example experiment

Trial run order	Standard order number	Adhesive type	Conductor material	Cure time	I.C. post coating	Bond strength (y)
6	1	D2A	copper	90	tin	73
4	2	D2A	copper	120	silver	88
2	3	D2A	nickel	90	silver	81
7	4	D2A	nickel	120	tin	77
3	5	H-1-E	copper	90	silver	83
1	6	H-1-E	copper	120	tin	81
8	7	H-1-E	nickel	90	tin	74
5	8	H-1-E	nickel	120	silver	90

FIGURE 4.6 Report form for example experiment

Random Order Trial Number	Standard Order Trial Number	Response Observed Values y	A: ADHESIVE TYPE		B: CONDUCTOR MATERIAL		C: CURE TIME		A x B		A x C		B x C		D: I.C. POST COATING	
			D2A	HIE	Cu	Ni	90 MIN	120 MIN							TIN	SILVER
			1	2	1	2	1	2	1	2	1	2	1	2	1	2
6	1	73	73		73		73			73		73		73	73	
4	2	88	88		88			88		88	88		88			88
2	3	81	81			81	81		81			81	81			81
7	4	77	77			77		77	77		77			77	77	
3	5	83		83	83		83		83		83			83		83
1	6	81		81	81			81	81			81	81		81	
8	7	74		74		74	74			74	74		74		74	
5	8	90		90		90		90		90		90		90		90
TOTAL		647	319	328	325	322	311	336	322	325	322	325	324	323	305	342
NUMBER OF VALUES		8	4	4	4	4	4	4	4	4	4	4	4	4	4	4
AVERAGE		80.9	79.8	82.0	81.3	80.5	77.8	84.0	80.5	81.3	80.5	81.3	81.0	80.8	76.3	85.5
EFFECT			2.2		-0.8		6.2		0.8		0.8		-0.2		9.2	

FIGURE 4.7 Response table for example

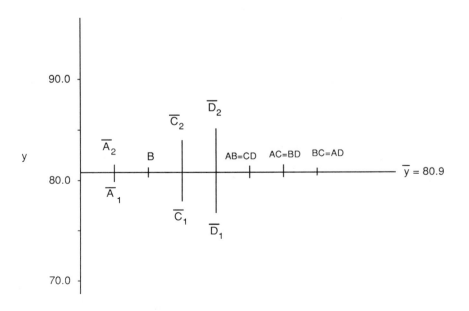

FIGURE 4.8 Graph of estimated effects

factor A (adhesive type) affect the response variable (bond strength). Factor B and all interaction terms appear to have no measurable effect on the average response. (To avoid cluttering up the figure, levels for B and the interaction terms are not specifically identified in Figure 4.8). Based on Figure 4.8, both C and D should be set at their high levels (cure time at 120 minutes and silver as the I.C. post coating) in order to maximize bond strength. The H-1-E adhesive may be slightly better than the D2A adhesive, but the estimated difference is slight. If adhesive H-1-E were more expensive to use, the possible benefits from its use may not be worth the added cost. As always, confirmatory trials should be run at the selected factor levels.

Example 4.2

An injection molding process produces structural film panels. The degree of flatness of these sheets is a critical quality characteristic. Four factors were thought to affect flatness: melt temperature, mold temperature, cure time, and injection speed. An eight-run experiment was performed using the levels for these factors listed in Figure 4.9. Figure 4.10 gives the completed report form for the experiment, with trials listed in the (random) order in which they were performed. The response variable measures flatness in thousandths of an inch. Smaller values indicate a flatter surface. The completed response table for the experiment is shown in Figure 4.11.

Based on the estimates at the bottom of Figure 4.11, it appears that factors A and B have relatively large effects on the response. The AB column pro-

Factor	Low level	High level
A. Melt temperature	500° F	550° F
B. Mold temperature	80° F	140° F
C. Cure time	150 sec	200 sec
D. Injection speed	1.00 sec	2.25 sec

FIGURE 4.9 Factor levels for example experiment

Trial run order	Standard order number	Melt temperature	Mold temperature	Cure time	Injection speed	Flatness (y)
1	3	500	140	150	2.25	46
2	1	500	80	150	1.00	54
3	6	550	80	200	1.00	45
4	4	500	140	200	1.00	50
5	2	500	80	200	2.25	55
6	7	550	140	150	1.00	30
7	5	550	80	150	2.25	46
8	8	550	140	200	2.25	24

FIGURE 4.10 Report form for example experiment

Random Order Trial Number	Standard Order Trial Number	Response Observed Values y	A: MELT TEMPERATURE 500°F (1)	A: MELT TEMPERATURE 550°F (2)	B: MOLD TEMPERATURE 80°F (1)	B: MOLD TEMPERATURE 140°F (2)	C: CURE TIME 150 SEC (1)	C: CURE TIME 200 SEC (2)	A × B (1)	A × B (2)	A × C (1)	A × C (2)	B × C (1)	B × C (2)	D: INJECTION SPEED 1.00 SEC (1)	D: INJECTION SPEED 2.25 SEC (2)
2	1	54	54		54		54			54		54		54	54	
5	2	55	55		55			55		55	55		55			55
1	3	46	46			46	46		46			46	46			46
4	4	50	50			50		50	50		50			50	50	
7	5	46		46	46		46		46		46			46		46
3	6	45		45	45			45	45			45	45		45	
6	7	30		30		30	30			30	30		30		30	
8	8	24		24		24		24		24		24		24		24
TOTAL		350	205	145	200	150	176	174	187	163	181	169	176	174	179	171
NUMBER OF VALUES		8	4	4	4	4	4	4	4	4	4	4	4	4	4	4
AVERAGE		43.8	51.3	36.3	50.0	37.5	44.0	43.5	46.8	40.8	45.3	42.3	44.0	43.5	44.8	42.8
EFFECT			−15.0		−12.5		−0.5		−6.0		−3.0		−0.5		−2.0	

FIGURE 4.11 Response table for example

vides the next largest estimated effect. But is this possible effect due to an AB interaction or its alias, the CD interaction? In this case, since A and B have large estimated main effects, while both C and D have small estimated main effects, it is reasonable tentatively to attribute this interaction effect, if it is real, to the AB interaction. In Figure 4.12 the average response is calculated for each combination of levels of factors A and B. These average values are plotted in Figure 4.13. Since the goal here is to minimize the average response, it appears that factors A and B should both be set at their high levels, since the average response is lowest for that combination of factor levels. Additional runs should of course be made to confirm this decision.

Five factors in eight trials

Suppose now that we wanted to evaluate the effects of *five* factors in eight experimental runs. We would again start with the experiment described in Figure 4.1, and would assign the two added factors, D and E, to two of the last four columns in Figure 4.1. What columns in Figure 4.1 should be assigned to factors D and E? For the four-factor experiment, D was assigned to ABC. Suppose we again set D = ABC, and then set E = AB. What sort of alias structure will result? The aliases for main effects and two-factor interactions are:

A = BE, B = AE, C = DE, D = CE, E = AB = CD, AC = BD, AD = BC.

		B: Mold Temperature	
		1: 80° F	2: 140° F
A: Melt Temperature	1: 500° F	$54 + 55 = 109$ $\overline{A_1B_1} = \dfrac{109}{2} = 54.5$	$46 + 50 = 96$ $\overline{A_1B_2} = \dfrac{96}{2} = 48.0$
	2: 550° F	$46 + 45 = 91$ $\overline{A_2B_1} = \dfrac{91}{2} = 45.5$	$30 + 24 = 54$ $\overline{A_2B_2} = \dfrac{54}{2} = 27.0$

FIGURE 4.12 Average response at different levels of melt and mold temperatures

(These confounding relationships are obtained by treating the expressions $D = ABC$ and $E = AB$ as algebraic equations. Clearly they aren't really algebraic equations, but it can be proven that this "illusion" correctly identifies the alias structure of the design.) Note that main factors are now confounded with two-factor interactions. For the four-factor design in Figure 4.2, main effects were only confounded with three-factor interactions. Suppose we tried

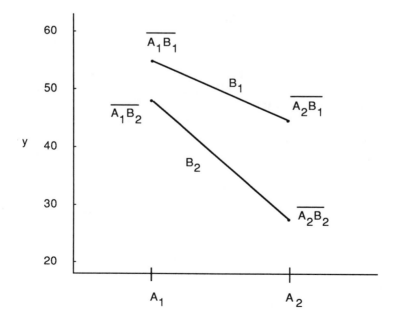

FIGURE 4.13 Graph of effect of AB interaction

| *Standard* | MAIN EFFECTS | | | | |
order	A	B	C	D	E
1	−1	−1	−1	1	1
2	−1	−1	1	1	−1
3	−1	1	−1	−1	1
4	−1	1	1	−1	−1
5	1	−1	−1	−1	−1
6	1	−1	1	−1	1
7	1	1	−1	1	−1
8	1	1	1	1	1

FIGURE 4.14 Design matrix for a five-factor experiment in eight trials

a different pair of design generators, say $D = AB$ and $E = AC$. This would produce a design with the following alias structure for main effects and two-factor interactions:

$$A = BD = CE, \ B = AD, \ C = AE, \ D = AB, \ E = AC, \ BC = DE, \ BE = CD.$$

Again the same basic result. In fact any attempt at analyzing five factors with eight trials will always produce confounding of main effects and two-factor interactions. For five factors in eight runs we will, in this book, assign D to the AB column and E to the AC column, since this is basically what is done by Taguchi and also by Western statisticians. The design matrix for this experiment is given in Figure 4.14.

Resolution of fractional factorial designs

The level of confounding of an experiment is called its *resolution*:

A *resolution III design* does not have any main effect aliased with other main effects, but main effects are aliased with two-factor interactions. Two-factor interactions may also be aliased with each other. The five-factor designs just discussed are of resolution III.

A *resolution IV design* does not have any main effects aliased with each other or with two-factor interactions. However, some two-factor interactions are aliased with other two-factor interactions. The four-factor design in Figure 4.2 is of resolution IV.

A *resolution V design* does not have any main effects or two-factor interaction effects aliased with each other, but some two-factor interactions are aliased with three-factor interactions. (See Box and Hunter, 1961, Part II, for further discussion on resolution V designs.)

Up to seven factors can be analyzed, at two levels each, with an eight-run experiment. But with five or more factors, the design will be of resolution III.

Resolution IV designs are used frequently because they seem to provide a good balance of useful information versus number of trials required. Resolution III designs provide a great deal of information for the small number of trials performed, but can be misleading if there is heavy confounding of factors. However, if a large number of factors must be included in an engineering study, it often makes sense to perform a resolution III design first, and then after the unimportant factors have been identified and removed from the study, additional resolution IV experiments can be performed with the remaining factors. Unless experimental runs are relatively inexpensive or the number of experimental factors is small or several interaction terms are critical to the study, resolution V experiments are probably too expensive for many situations.

Example 4.3

An engineer is interested in improving the efficiency of a deburring operation. The deburring machine uses wire brushes for material removal. Five factors are identified as affecting the rate of material removal. They are listed in Figure 4.15 and illustrated in Figure 4.16.

The design used for the experiment is given in Figure 4.14. As we saw earlier in this section, this will result in the following alias structure:

$A = BD = CE$, $B = AD$, $C = AE$, $D = AB$, $E = AC$, $BC = DE$, and $BE = CD$.

The completed response table for the experiment is given in Figure 4.17. The response variable is rate of material removal measured in cubic inches $\times$ 10^{-7} per revolution. The estimated effects are plotted in Figure 4.18. Based on Figure 4.18, factors A, B, and E appear significantly to affect the average response. But what about possible two-factor interactions? If two-factor interactions were present, we would expect to see large estimated two-factor effects between A and B, A and E, or B and E, since these were the factors with large main effects. However, none of these three pairs of factors has a large estimated interaction effect. So the data suggest that the three large estimated effects are only main factor effects. In order to maximize material removal, the deburring machine should be run at low levels for factors A, B, and E. That is, run the machine with:

- penetration depth of 0.12 inches,
- brush width 1.5 inches, and
- filament diameter of 0.010 inches.

Factor	Low level	High level
A. Penetration depth	0.12 in	0.17 in
B. Brush width	1.5 in	2.0 in
C. Number of filaments	20,000	25,000
D. Filament length	1.0 in	2.0 in
E. Filament diameter	0.010 in	0.015 in

FIGURE 4.15 Experimental levels for five-factor example

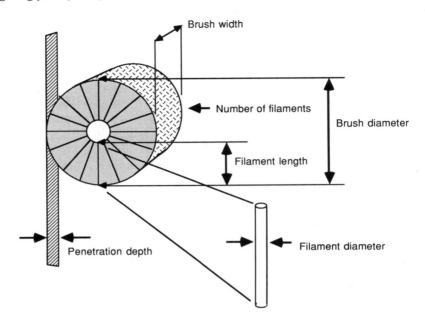

FIGURE 4.16 Deburring brush

Random Order Trial Number	Standard Order Trial Number	Response Observed Values y	A: BRUSH PENETRATION DEPTH 0.12 IN. (1)	0.17 IN. (2)	B: BRUSH WIDTH 1.5 IN. (1)	2.0 IN. (2)	C: NUMBER OF WIRE FILAMENTS 20k (1)	25k (2)	D: FILAMENT LENGTH 1.0 IN. (1)	2.0 IN. (2)	E: FILAMENT DIAMETER 0.010 IN. (1)	0.015 IN. (2)	BC (1)	BC (2)	CD (1)	CD (2)
5	1	123	123		123		123			123		123	123		123	
2	2	163	163		163			163		163	163		163			163
4	3	115	115			115	115		115			115	115			115
1	4	126	126			126		126	126		126			126	126	
6	5	150		150	150		150		150		150			150		150
7	6	114		114	114			114	114			114	114		114	
3	7	105		105		105	105			105	105		105		105	
8	8	76		76		76		76		76		76		76		76
TOTAL		972	527	445	550	422	493	479	505	467	544	428	497	475	468	504
NUMBER OF VALUES		8	4	4	4	4	4	4	4	4	4	4	4	4	4	4
AVERAGE		121.5	131.8	111.3	132.5	105.5	123.3	119.8	126.3	116.8	136.0	107.0	124.3	118.8	117.0	126.0
EFFECT			−20.5		−32.0		−3.5		−9.5		−29.0		−5.5		9.0	

FIGURE 4.17 Response table for five-factor example

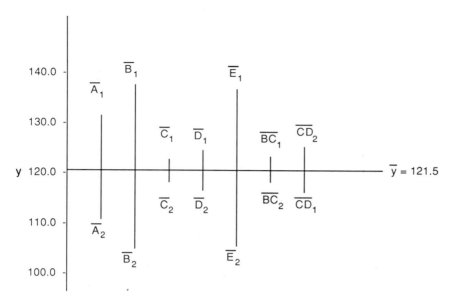

FIGURE 4.18 Graph of estimated effects for five-factor example

The estimated average material removal rate at these factor settings is:

$$\hat{y} = \text{(grand mean)} + \text{(A contribution)} + \text{(B contribution)} + \text{(E contribution)}$$
$$= 121.5 + (131.8 - 121.5) + (137.5 - 121.5) + (136.0 - 121.5)$$
$$= 121.5 + 10.3 + 16.0 + 14.5$$
$$= 162.3 \text{ in}^3 \times 10^{-7} \text{ per revolution.}$$

(Note: "$\hat{y}$" will be used to denote estimated average response for a given combination of factor levels.)

The settings suggested for factors A, B, and E should not, of course, be blindly set based on this one experiment. Additional experiments should be performed to confirm that these decisions were correct.

k factors in eight trials

Experiments for analyzing the effects of five, six or seven factors, at two levels each, using eight trials, are also based on Figure 4.1. Each factor is assigned one of the columns in Figure 4.1. Figure 4.1 is reproduced in Figure 4.19 with the -1s replaced by 1s, the 1s replaced by 2s, and the columns numbered 1 through 7, from left to right. Assignments of factors to the design matrix columns in Figure 4.19 for three, four, five, six, and seven factors are given below. The alias relationships for main effects and two-factor interactions are also provided. As an example of "reading" the information, for four factors in eight runs the experimenter should determine levels of factors A, B, C, and D for each run based on the values in columns 1, 2, 3, and 7, respectively. If this is done, then the AB and CD interaction effects will be confounded, as

Run no.	FACTOR COLUMNS						
	1	*2*	*3*	*4*	*5*	*6*	*7*
1	1	1	1	2	2	2	1
2	1	1	2	2	1	1	2
3	1	2	1	1	2	1	2
4	1	2	2	1	1	2	1
5	2	1	1	1	1	2	2
6	2	1	2	1	2	1	1
7	2	2	1	2	1	1	1
8	2	2	2	2	2	2	2

FIGURE 4.19 Basic design matrix for eight-run, two-level experiments

will the AC and BD interaction effects and the AD and BC interaction effects. All the designs used in this section are orthogonal. This means that, for example, a large B effect or large AB interaction effect will not distort the estimated A effect. But orthogonality does not help us sort out confounded effects such as AB and CD.

For three factors, assign:
    A to column 1
    B to column 2
    C to column 3

      Then:
          AB is in column 4
          AC is in column 5
          BC is in column 6

For four factors, assign:
    A to column 1
    B to column 2
    C to column 3
    D to column 7

      Then:
          AB = CD is in column 4
          AC = BD is in column 5
          AD = BC is in column 6

For five factors, assign:
    A to column 1
    B to column 2
    C to column 3
    D to column 4
    E to column 5

      Then:
          A = BD = CE is in column 1
          B = AD is in column 2

    C = AE is in column 3
    D = AB is in column 4
    E = AC is in column 5
    BC = DE is in column 6
    BE = CD is in column 7

For six factors, assign:
    A to column 1
    B to column 2
    C to column 3
    D to column 4
    E to column 5
    F to column 6

      Then:
        A = BD = CE is in column 1
        B = AD = CF is in column 2
        C = AE = BF is in column 3
        D = AB = EF is in column 4
        E = AC = DF is in column 5
        F = BC = DE is in column 6
        AF = BE = CD is in column 7

For 7 factors, assign:
    A to column 1
    B to column 2
    C to column 3
    D to column 4
    E to column 5
    F to column 6
    G to column 7

      Then:
        A = BD = CE = FG is in column 1
        B = AD = CF = EG is in column 2
        C = AE = BF = DG is in column 3
        D = AB = EF = CG is in column 4
        E = AC = BG = DF is in column 5
        F = AG = BC = DE is in column 6
        G = AF = BE = CD is in column 7

Example 4.4

In a robotic grinding process, vibration was a chronic problem. A project team set up to deal with the problem identified seven potential factors which could affect the amount of vibration: grinding tool diameter, grinding tool length, tool material grit size, tool material structure, tool holder preload, and tool feed rate. The project team selected two levels for each factor at which to conduct experimental trials. They are listed in Figure 4.20. Because of setup and run time costs, the team decided to use an eight-run experiment. Michael,

Factor	Low level	High level
A. Grinding tool diameter	1.0 in	1.5 in
B. Grinding tool length	1.0 in	2.0 in
C. Tool material grit size	80/in	120/in
D. Tool material structure	1 oz/in^3	4 oz/in^3
E. Tool holder preload	1 lb	4 lb
F. Tool rpm	15,000 rpm	20,000 rpm
G. Tool feed rate	2 in/min	4 in/min

FIGURE 4.20 Experimental factor levels

an engineer who had analyzed a similar problem earlier, told the team that interaction effects among the factors should be negligible.

The data from the experiment are recorded in column 3 of Figure 4.21. The factor effects are also estimated in that response table. The estimated effects are presented graphically in Figure 4.22. The object of the study is to reduce the level of vibration (smaller-the-better). Figure 4.22 indicates that the high level for factor A (grinding tool diameter) and the low levels for factors C (grit size) and E (tool holder preload) should be used. Setting factors B, D, and F at their high levels should also be considered, although the "effects" observed here may represent random variation rather than real effects. Factor G (tool feed rate) should be set at its high level since feed rate does not appear to affect vibration, and productivity can be increased by operating at the higher

Random Order Trial Number	Standard Order Trial Number	Response Observed Values y	A: Grinding Tool Diameter 1.0 IN. (1)	1.5 IN. (2)	B: Grinding Tool Length 1.0 IN. (1)	2.0 IN. (2)	C: Tool Material Grit Size 80 (1)	120 (2)	D: Tool Material Structure 1 oz. (1)	4 oz. (2)	E: Tool Holder Preload 1 lb. (1)	4 lb. (2)	F: Tool R.P.M. 15,000 (1)	20,000 (2)	G: Tool Feed 2 IN./MIN (1)	4 IN./MIN (2)
1		77.4	77.4	77.4	77.4		77.4			77.4		77.4		77.4	77.4	
2		68.3	68.3		68.3			68.3		68.3	68.3		68.3			68.3
3		81.9	81.9			81.9	81.9		81.9			81.9	81.9			81.9
4		66.2	66.2			66.2		66.2	66.2		66.2			66.2	66.2	
5		42.1		42.1	42.1		42.1		42.1		42.1			42.1		42.1
6		78.3		78.3	78.3			78.3	78.3			78.3	78.3		78.3	
7		39.0		39.0		39.0	39.0			39.0	39.0		39.0		39.0	
8		68.4		68.4		68.4		68.4		68.4		68.4		68.4		68.4
TOTAL		521.6	293.8	227.8	266.1	255.5	240.4	281.2	268.5	253.1	215.6	306.0	267.5	254.1	260.9	260.7
NUMBER OF VALUES		8	4	4	4	4	4	4	4	4	4	4	4	4	4	4
AVERAGE		65.20	73.45	56.95	66.53	63.88	60.10	70.30	67.13	63.28	53.90	76.50	66.88	63.53	65.23	65.18
EFFECT				-16.50		-2.65		10.20		-3.85		22.60		-3.35		-0.05

FIGURE 4.21 Completed response table for example data

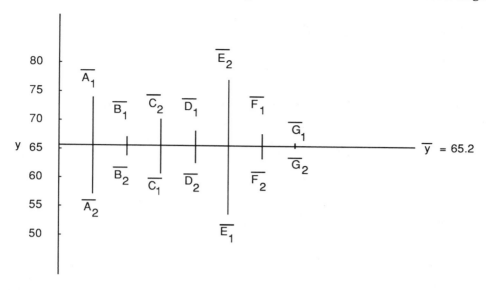

FIGURE 4.22 Estimated effects from example

feed rate. Given the limited number of trials used in the experiment, and the lack of dramatically large estimated effects, confirmatory runs should be made as soon as possible. There is also the concern here that it was assumed, based on engineering judgment, that there were no interaction effects. If possible interactions were considered, the conclusions of the analysis might be different. For example, since the main E effect is confounded with the AC interaction effect, are we really seeing a significant AC interaction effect in the analysis?

Example 4.5

Welded steel joints produced by a metal inert gas process were found to be unreliable. Seven factors have been identified as possibly affecting the strength of the welds. It was decided to conduct an eight-run experiment to explore further the effects these factors have on weld strength. The seven factors, and the experimental levels for each, are given in Figure 4.23. Figure 4.24 is the response table for the experiment.

Factor	Low level	High level
A. Raw material alloy content	low	high
B. Wire electrode diameter	0.030 in	0.035 in
C. Weld circuit current	50 amps	180 amps
D. Argon carbon dioxide shielding gas	low CO_2	high CO_2
E. Welding voltage	15 volts	25 volts
F. Wire feed speed	100 in/min	300 in/min
G. Preheat	no	yes

FIGURE 4.23 Experimental factors and levels

Random Order Trial Number	Standard Order Trial Number	Response Observed Values y	A: Raw material alloy content		B: wire electrode diameter		C: weld circuit current		D: Argon carbon dioxide		E: welding voltage		F: wire feed speed		G: Preheat	
			low 1	high 2	1	2	1	2	1	2	1	2	1	2	1	2
	1	147.2	147.2		147.2		147.2			147.2		147.2		147.2	147.2	
	2	91.3		91.3	91.3		91.3			91.3	91.3		91.3			91.3
	3	72.7	72.7			72.7	72.7		72.7			72.7	72.7			72.7
	4	87.4		87.4		87.4	87.4		87.4		87.4			87.4	87.4	
	5	84.1	84.1		84.1			84.1	84.1		84.1			84.1		84.1
	6	78.2		78.2	78.2			78.2	78.2			78.2	78.2		78.2	
	7	94.6	94.6			94.6		94.6		94.6	94.6		94.6		94.6	
	8	138.8		138.8		138.8		138.8		138.8		138.8		138.8		138.8
TOTAL		794.3	398.6	395.7	400.8	393.5	398.6	395.7	322.4	471.9	357.4	436.9	336.8	457.5	407.4	386.9
NUMBER OF VALUES		8	4	4	4	4	4	4	4	4	4	4	4	4	4	4
AVERAGE		99.29	99.65	98.93	100.2	98.38	99.65	98.93	80.60	117.98	89.35	109.23	84.20	114.38	101.85	96.23
EFFECT				−0.72		−1.82		−0.72		37.38		19.88		30.18		−5.12

FIGURE 4.24 Response table for seven-factor, eight-run experiment

From Figure 4.24 we see that three estimated effects are much larger than the rest: D, E, and F. At first reading, it would appear that the three key factors, in order of estimated effect, are:

D: argon carbon dioxide shielding gas,

F: Wire feed speed, and

E: Welding voltage.

But if the alias structure for a seven-factor, eight-run experiment is considered, we find that main effects D, E, and F are confounded with two-factor interactions in the following ways:

$$D = AB = EF = CG,$$
$$E = AC = BG = DF,$$
$$F = AG = BC = DE.$$

Suppose we assume that since factors A, B, C, and G do not have significant main effects, they are not involved in any significant two-factor interactions either. Then the confounding relations just listed reduce to:

D = EF,

E = DF,

F = DE.

Do the levels of D, E, and F all affect weld strength, or do just two of them have real effects, and have a two-factor interaction as well? At this point we cannot tell. Additional experimentation is needed. Replicating the experiment will probably not help. We will just confirm the uncertain situation. There is, however, a special type of experimental design called a *fold-over design* which will help in dealing with situations like this. It is explained in the next section. At the end of that section we will return to our example and solve the mystery of which factors are really affecting weld strength.

4.2 Folding Over an Eight-run Experimental Design

When analyzing more than three factors in eight runs we sometimes get mixed signals in the form of confounded effects with large estimated values. Which of the confounded effects are affecting the response variable? A basic rule mentioned earlier is generally to assume two-factor interactions are zero if the factors involved in the interaction have small estimated main effects. This rule will not always resolve problems in interpreting effects, however. Suppose the design given in Figure 4.14 for five factors was used and columns A, B, and D produced large estimated effects. Factors C and E can be eliminated as contributors, but this still leaves the following alias structure: A = BD, B = AD, D = AB. Are factors A, B, and D all important, or are just two of them contributing through main effects and their interaction? We can't tell at this point. More experimentation is needed. The issue could perhaps be resolved by performing another eight-run experiment involving just these three factors. But it would be better to construct an experiment which would "fit" with the first experiment, so that the data from both experiments could be used in the estimates. This is done by performing a second eight-run experiment where the factor levels are all the opposite of what they were in the first experiment. That is, interchange the −1s and 1s in Figure 4.14 (or, more generally, the 1s and 2s in Figure 4.19) before carrying out the second experiment. When the two experiments are then combined, the resulting experiment will have resolution IV, even if the two experiments separately are resolution III. This is a form of what is called a *fold-over design*. The rationale behind fold-over designs is explained in Box and Hunter, 1961.

The factor matrix for the $8 + 8 = 16$ trials of a folded-over eight-run experiment is given in Figure 4.25 using the 1,2 notation. Note that the top half of Figure 4.25 is the same as Figure 4.19 except for the "*Block*" column. This column contains 1s for the first half of the experiment and 2s for the second half. The two halves of the experiment are called blocks. This is another experimental factor added to the experiment. The *Block* column will be used to estimate the "block effect," which is the difference between the average re-

Run		FACTOR COLUMNS						
no.	Block	1	2	3	4	5	6	7
1	1	1	1	1	2	2	2	1
2	1	1	1	2	2	1	1	2
3	1	1	2	1	1	2	1	2
4	1	1	2	2	1	1	2	1
5	1	2	1	1	1	1	2	2
6	1	2	1	2	1	2	1	1
7	1	2	2	1	2	1	1	1
8	1	2	2	2	2	2	2	2
9	2	2	2	2	1	1	1	2
10	2	2	2	1	1	2	2	1
11	2	2	1	2	2	1	2	1
12	2	2	1	1	2	2	1	2
13	2	1	2	2	2	2	1	1
14	2	1	2	1	2	1	2	2
15	2	1	1	2	1	2	2	2
16	2	1	1	1	1	1	1	1

FIGURE 4.25 Design matrix for sixteen-run fold-over experiment

sponses for the two halves of the experiment. (Block effects will be discussed further in section '4.5.) The order of performing the runs should, of course, be randomized within each block.

There are three steps to performing a fold-over experiment:

1. Use the standard response table for eight-run experiments (Figure 3.25) for the first block of eight trials. Calculate estimated effects as usual.
2. For the second, folded-over, eight-run experiment, use the response table in Figure 3.25 in one of the following two ways:
   a. For each pair of columns for factor effects, label the left column "1" and right column "2". Write response values in the shaded areas of the effect columns, instead of the clear, underlined areas (see Figure 4.28, for example), *or*
   b. Reverse the "1" and "2" headings of the effects columns, putting the "2" column to the left of the "1" column, and write the observed response values in the clear, underlined areas as usual.
   With either method, calculate estimated effects by subtracting the average in column "1" from the average in column "2".
3. The estimated effect values from the two response tables are now used to estimate factor main effects *free of interaction effects* and also to estimate two-factor interactions. Write the estimated effects from steps 1 and 2 above in columns 2 and 3 of the form found in Figure 4.26. For each row in Figure 4.26, calculate the values for columns 4 and 5 using the estimates recorded in columns 2 and 3.

The effects estimated in the last two columns of Figure 4.26 are summarized in Figure 4.27 for five, six, and seven factors. For three or four factors, folding

Column	Experiment 1 Effect E_1	Experiment 2 Effect E_2	Main Effect $\dfrac{E_1 + E_2}{2}$	Interaction Effect $\dfrac{E_1 - E_2}{2}$
Y*				
1				
2				
3				
4				
5				
6				
7				

(* Note: for the "Y" column, E_1 and E_2 are the average values from the two response tables.)

FIGURE 4.26 Fold-over design effects calculation table

Row of Fig. 4.26	Number of Factors					
	5		6		7	
	Effect Column		Effect Column		Effect Column	
	Main	Interaction	Main	Interaction	Main	Interaction
Y	average	block	average	block	average	block
1	A	BD=CE	A	BD=CE	A	BD=CE=FG
2	B	AD	B	AD=CF	B	AD=CF=EG
3	C	AE	C	AE=BF	C	AE=BF=DG
4	D	AB	D	AB=EF	D	AB=CG=EF
5	E	AC	E	AC=DF	E	AC=BG=DF
6	*	BC=DE	F	BC=DE	F	AG=BC=DE
7	*	BE=CD	*	AF=BE=CD	G	AF=BE=CD

(Note: * denotes three-factor effects.)

FIGURE 4.27 Effects estimated using Figure 4.26

over is usually unnecessary since main effects are not confounded with two-factor interactions. Note in Figure 4.27 that for five factors, the two-factor interactions involving factor A are not confounded with other effects. So if there is one factor in an experiment which is felt to be most important, or is most likely to interact with other factors, it should be made factor A *before the experiment begins*. Figure 4.27 does not list three-factor or higher-level interactions. Experimenters often assume that three- and four-factor interactions are insignificant in comparison to main effects and two-factor interactions. Note in Figure 4.27 that for five and six factors there are three cases of estimates of three-factor interactions. These estimates should be included in the normal plot of effects since they often represent approximate normal variability, and so provide a reference base for comparison with other estimated effects. For further discussion on fold-over designs see Box, Hunter, and Hunter (1978).

Example 4.6

Section 4.1 ended with an eight-run experiment in seven factors (see Figures 4.23 and 4.24). Three main effects were confounded with two-factor interactions in such a way that it was impossible to determine whether three main effects were affecting the response, or if two of the three factors and their interaction were most important. This sort of situation calls for a fold-over experiment. The completed response table from the folded-over experiment is given in Figure 4.28. Note that the observed response values are written in the shaded areas of the response table, since the levels of factors are the opposite of what they were during the first eight trials. The estimated effects from the first eight runs (Figure 4.24) and second eight runs (Figure 4.28) are combined in Figure 4.29 to produce estimates of all main and two-factor interactions. Fifteen of the sixteen estimates in the last two columns of Figure 4.29 are plotted on normal probability paper in Figure 4.30. The "main effect" estimate from the "y" row is not plotted since it is assumed to be real, and its inclusion would result in scaling problems for the graph. From inspection of Figures 4.29 and 4.30, it is clear that the following three effects significantly affect the response:

- main effect row 5, which is factor E (see Figure 4.27),
- main effect row 6, which is factor F, and
- interaction effect row 4, which is AB=CG=EF.

Since E and F are the only significant main effects, we tentatively conclude, subject to confirmatory experiments, that the interaction effect in row 4 is most likely the EF interaction.

Recall that after the first eight runs, it appeared that factor D might be significant. Had we not separated the main effects from the interaction effects, we might have made a serious mistake here by assuming the D effect was real.

The average responses at the four different combinations of levels for factors E and F are given in Figure 4.31 and plotted in Figure 4.32. Since the

Random Order Trial Number	Standard Order Trial Number	Response Observed Values y	A:1	A:2	B:1	B:2	C:1	C:2	D:1	D:2	E:1	E:2	F:1	F:2	G:1	G:2
	1	89.8		89.8		89.8		89.8	89.8		89.8		89.8			89.8
	2	137.4		137.4		137.4	137.4		137.4			137.4		137.4	137.4	
	3	82.7		82.7	82.7			82.7		82.7	82.7			82.7	82.7	
	4	72.2		72.2	72.2		72.2			72.2		72.2	72.2			72.2
	5	71.9	71.9			71.9		71.9		71.9		71.9	71.9		71.9	
	6	87.3	87.3			87.3	87.3			87.3	87.3			87.3		87.3
	7	144.6	144.6		144.6			144.6	144.6			144.6		144.6		144.6
	8	93.2	93.2		93.2		93.2		93.2		93.2		93.2		93.2	
TOTAL		779.1	397.0	382.1	392.7	386.4	390.1	389.0	465.0	314.1	353.0	426.1	327.1	452.0	385.2	393.9
NUMBER OF VALUES		8	4	4	4	4	4	4	4	4	4	4	4	4	4	4
AVERAGE		97.39	99.25	95.53	98.18	96.60	97.53	97.25	116.25	78.53	88.25	106.53	81.78	113.00	96.30	98.48
EFFECT			-3.72		-1.58		-0.28		-37.72		18.28		31.22		2.18	

FIGURE 4.28 Response table for a fold-over experiment

Column	Experiment 1 Effect E_1	Experiment 2 Effect E_2	Main Effect $\dfrac{E_1+E_2}{2}$	Interaction Effect $\dfrac{E_1-E_2}{2}$
Y*	99.29	97.39	98.34	0.95
1	-0.72	-3.72	-2.22	1.50
2	-1.82	-1.58	-1.70	-0.12
3	-0.72	-0.28	-0.50	-0.22
4	37.38	-37.72	-0.17	37.55
5	19.88	18.28	19.08	0.80
6	30.18	31.22	30.70	-0.52
7	-5.12	2.18	-1.47	-3.65

FIGURE 4.29 Combined example estimates

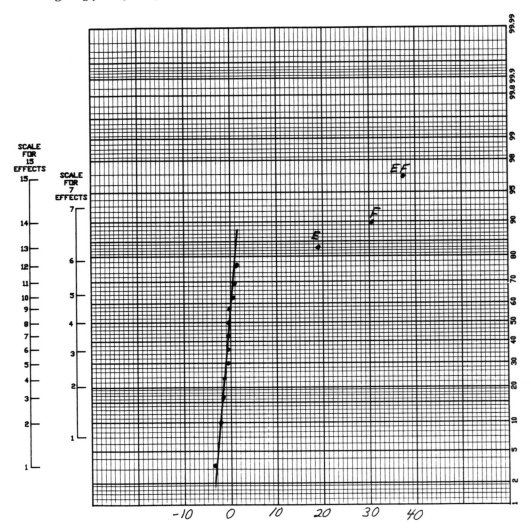

FIGURE 4.30 Normal plot for fold-over effects

goal is to increase weld strength (larger-the-better), both E and F should be set at their high levels. That combination produces the largest average response (142).

4.3 Fractional Factorial Designs in Sixteen Runs

Section 3.9 explained how to analyze the effects of four factors in sixteen runs. It is possible to analyze the effects of up to fifteen factors in sixteen trials, if we are willing to tolerate some confounding of effects. The levels

		E: Welding Voltage	
		1: 15 Volts	2: 25 Volts
F. Wire Feed Speed	1: 100 in./ minute	91.3 94.6 89.8 93.2 ——— 368.9 $\overline{E_1F_1} = \dfrac{368.9}{4} = 92.23$	72.7 78.2 72.2 71.9 ——— 295.0 $\overline{E_2F_1} = \dfrac{295.0}{4} = 73.75$
	2: 700 in./ minute	87.4 84.1 82.7 87.3 ——— 341.5 $\overline{E_1F_2} = \dfrac{341.5}{4} = 85.38$	147.2 138.8 137.4 144.6 ——— 568.0 $\overline{E_2F_2} = \dfrac{568.0}{4} = 142.0$

FIGURE 4.31 Average response values for different levels of factors E and F

assigned to the various factors are determined by the values listed in Figure 3.53. Figure 3.53 is reproduced, with −1s replaced by 1s and 1s replaced by 2s, in Figure 4.33. The columns have been labeled 1, 2, 3, . . . , 15 in Figure 4.33 to simplify assignment of factors to the columns of the matrix. Recommended assignments of factors to the columns of Figure 4.33, and the resul-

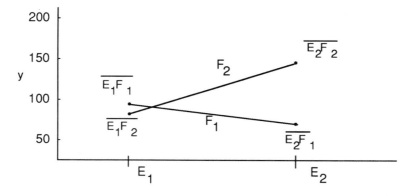

FIGURE 4.32 Plot of average response for different levels of factors E and F

Standard Order	Column Number														
	1	2	3	4	5	6	7	8	9	10	11	12	13	14	15
1	1	1	1	1	2	2	2	2	2	2	1	1	1	1	2
2	1	1	1	2	2	2	1	2	1	1	1	2	2	2	1
3	1	1	2	1	2	1	2	1	2	1	2	1	2	2	1
4	1	1	2	2	2	1	1	1	1	2	2	2	1	1	2
5	1	2	1	1	1	2	2	1	1	2	2	2	1	2	1
6	1	2	1	2	1	2	1	1	2	1	2	1	2	1	2
7	1	2	2	1	1	1	2	2	1	1	1	2	2	1	2
8	1	2	2	2	1	1	1	2	2	2	1	1	1	2	1
9	2	1	1	1	1	1	1	2	2	2	2	2	2	1	1
10	2	1	1	2	1	1	2	2	1	1	2	1	1	2	2
11	2	1	2	1	1	2	1	1	2	1	1	2	1	2	2
12	2	1	2	2	1	2	2	1	1	2	1	1	2	1	1
13	2	2	1	1	2	1	1	1	1	2	1	1	2	2	2
14	2	2	1	2	2	1	2	1	2	1	1	2	1	1	1
15	2	2	2	1	2	2	1	2	1	1	2	1	1	1	1
16	2	2	2	2	2	2	2	2	2	2	2	2	2	2	2

FIGURE 4.33 Design matrix for sixteen-run experiment

tant alias relationships among main effects and two-factor interactions, are given below for four through fifteen factors.

For four factors, assign:
    A to column 1
    B to column 2
    C to column 3
    D to column 4

      Then:
          AB is in column 5
          AC is in column 6
          AD is in column 7
          BC is in column 8
          BD is in column 9
          CD is in column 10
          Columns 11–15 contain three- and four-factor interactions

For five factors, assign:
    A to column 1
    B to column 2
    C to column 3

D to column 4
E to column 15

    Then:
        AB is in column 5
        AC is in column 6
        AD is in column 7
        BC is in column 8
        BD is in column 9
        CD is in column 10
        DE is in column 11
        CE is in column 12
        BE is in column 13
        AE is in column 14

For 6 factors, assign:
    A to column 1
    B to column 2
    C to column 3
    D to column 4
    E to column 11
    F to column 12

    Then:
        AB = CE = DF is in column 5
        AC = BE is in column 6
        AD = BF is in column 7
        BC = AE is in column 8
        BD = AF is in column 9
        CD = EF is in column 10
        CF = DE is in column 15
        Three-factor interactions are in columns 13 and 14

For seven factors, assign:
    A to column 1
    B to column 2
    C to column 3
    D to column 4
    E to column 11
    F to column 12
    G to column 13

    Then:
        AB = CE = DF is in column 5
        AC = BE = DG is in column 6
        AD = BF = CG is in column 7
        AE = BC = FG is in column 8
        AF = BD = EG is in column 9
        AG = CD = EF is in column 10
        BG = CF = DE is in column 15
        Three-factor interactions are in column 14

For eight factors, assign:
    A to column 1
    B to column 2
    C to column 3
    D to column 4
    E to column 11
    F to column 12
    G to column 13
    H to column 14

      Then:
          $AB = CE = DF = GH$ is in column 5
          $AC = BE = DG = FH$ is in column 6
          $AD = BF = CG = EH$ is in column 7
          $AE = BC = DH = FG$ is in column 8
          $AF = BD = CH = EG$ is in column 9
          $AG = BH = CD = EF$ is in column 10
          $AH = BG = CF = DE$ is in column 15

For nine factors, assign:
    A to column 1
    B to column 2
    C to column 3
    D to column 4
    E to column 11
    F to column 12
    G to column 13
    H to column 14
    I to column 15

      Then:
          $A = HI$ is in column 1
          $B = GI$ is in column 2
          $C = FI$ is in column 3
          $D = EI$ is in column 4
          $AB = CE = DF = GH$ is in column 5
          $AC = BE = DG = FH$ is in column 6
          $AD = BF = CG = EH$ is in column 7
          $AE = BC = DH = FG$ is in column 8
          $AF = BD = CH = EG$ is in column 9
          $AG = BH = CD = EF$ is in column 10
          $E = DI$ is in column 11
          $F = CI$ is in column 12
          $G = BI$ is in column 13
          $H = AI$ is in column 14
          $I = AH = BG = CF = DE$ is in column 15

For ten factors, assign:
    A to column 1
    B to column 2

C to column 3
D to column 4
E to column 11
F to column 12
G to column 13
H to column 14
I to column 15
J to column 5

    Then:

        $A = BJ = HI$ is in column 1
        $B = AJ = GI$ is in column 2
        $C = EJ = FI$ is in column 3
        $D = EI = FJ$ is in column 4
        $J = AB = CE = DF = GH$ is in column 5
        $AC = BE = DG = FH$ is in column 6
        $AD = BF = CG = EH$ is in column 7
        $AE = BC = DH = FG$ is in column 8
        $AF = BD = CH = EG$ is in column 9
        $AG = BH = CD = EF = IJ$ is in column 10
        $E = CJ = DI$ is in column 11
        $F = CI = DJ$ is in column 12
        $G = BI = HJ$ is in column 13
        $H = AI = GJ$ is in column 14
        $I = AH = BG = CF = DE$ is in column 15

For eleven factors, assign:
    A to column 1
    B to column 2
    C to column 3
    D to column 4
    E to column 11
    F to column 12
    G to column 13
    H to column 14
    I to column 15
    J to column 5
    K to column 6

    Then:

        $A = BJ = CK = HI$ is in column 1
        $B = AJ = EK = GI$ is in column 2
        $C = AK = EJ = FI$ is in column 3
        $D = EI = FJ = GK$ is in column 4
        $J = AB = CE = DF = GH$ is in column 5
        $K = AC = BE = DG = FH$ is in column 6
        $AD = BF = CG = EH$ is in column 7
        $AE = BC = DH = FG = JK$ is in column 8
        $AF = BD = CH = EG = IK$ is in column 9

AG = BH = CD = EF = IJ is in column 10
E = BK = CJ = DI is in column 11
F = CI = DJ = HK is in column 12
G = BJ = DK = HJ is in column 13
H = AI = FK = GJ is in column 14
I = AH = BG = CF = DE is in column 15

For twelve factors, assign:

A to column 1
B to column 2
C to column 3
D to column 4
E to column 11
F to column 12
G to column 13
H to column 14
I to column 15
J to column 5
K to column 6
L to column 7

Then:

A = BJ = CK = DL = HI is in column 1
B = AJ = EK = FL = GI is in column 2
C = AK = EJ = FI = GL is in column 3
D = AL = EI = FJ = GK is in column 4
J = AB = CE = DF = GH is in column 5
K = AC = BE = DG = FH is in column 6
L = AD = BF = CG = EH is in column 7
AE = BC = DH = FG = IL = JK is in column 8
AF = BD = CH = EG = IK = JL is in column 9
AG = BH = CD = EF = IJ = KL is in column 10
E = BK = CJ = DI = HL is in column 11
F = BL = CI = DJ = HK is in column 12
G = BI = CL = DK = HJ is in column 13
H = AI = EL = FK = GJ is in column 14
I = AH = BG = CF = DE is in column 15

For thirteen factors, assign:

A to column 1
B to column 2
C to column 3
D to column 4
E to column 11
F to column 12
G to column 13
H to column 14
I to column 15
J to column 5
K to column 6

L to column 7
M to column 8

Then:

A = BJ = CK = DL = EM = HI is in column 1
B = AJ = CM = EK = FL = GI is in column 2
C = AK = BM = EJ = FI = GL is in column 3
D = AL = EI = FJ = GK = HM is in column 4
J = AB = CE = DF = GH = KM is in column 5
K = AC = BE = DG = FH = JM is in column 6
L = AD = BF = CG = EH = IM is in column 7
M = AE = BC = DH = FG = IL = JK is in column 8
AF = BD = CH = EG = IK = JL is in column 9
AG = BH = CD = EF = IJ = KL is in column 10
E = AM = BK = CJ = DI = HL is in column 11
F = BL = CI = DJ = GM = HK is in column 12
G = BI = CL = DK = FM = HJ is in column 13
H = AI = DM = EL = FK = GJ is in column 14
I = AH = BG = CF = DE = LM is in column 15

For fourteen factors, assign:

A to column 1
B to column 2
C to column 3
D to column 4
E to column 11
F to column 12
G to column 13
H to column 14
I to column 15
J to column 5
K to column 6
L to column 7
M to column 8
N to column 9

Then:

A = BJ = CK = DL = EM = FN = HI is in column 1
B = AJ = CM = DN = EK = FL = GL is in column 2
C = AK = BM = EJ = FI = GL = HN is in column 3
D = AL = BN = EI = FJ = GK = HM is in column 4
J = AB = CE = DF = GH = KM = LN is in column 5
K = AC = BE = DG = FH = IN = JM is in column 6
L = AD = BF = CG = EH = IM = JN is in column 7
M = AE = BC = DH = FG = IL = JK is in column 8
N = AF = BD = CH = EG = IK = JL is in column 9
AG = BH = CD = EF = IJ = KL = MN is in column 10
E = AM = BK = CJ = DI = GN = HL is in column 11
F = AN = BL = CI = DJ = GM = HK is in column 12
G = BI = CL = DK = EN = FM = HJ is in column 13

$$H = AI = CN = DM = EL = FK = GJ \text{ is in column } 14$$
$$I = AH = BG = CF = DE = KN = LM \text{ is in column } 15$$

For fifteen factors, assign:

A to column 1
B to column 2
C to column 3
D to column 4
E to column 11
F to column 12
G to column 13
H to column 14
I to column 15
J to column 5
K to column 6
L to column 7
M to column 8
N to column 9
O to column 10

Then:

$$A = BJ = CK = DL = EM = FN = GO = HI \text{ is in column } 1$$
$$B = AJ = CM = DN = EK = FL = GI = HO \text{ is in column } 2$$
$$C = AK = BM = DO = EJ = FI = GL = HN \text{ is in column } 3$$
$$D = AL = BN = CO = EI = FJ = GK = HM \text{ is in column } 4$$
$$J = AB = CE = DF = GH = IO = KM = LN \text{ is in column } 5$$
$$K = AC = BE = DG = FH = IN = JM = LO \text{ is in column } 6$$
$$L = AD = BF = CG = EH = IM = JN = KO \text{ is in column } 7$$
$$M = AE = BC = DH = FG = IL = JK = NO \text{ is in column } 8$$
$$N = AF = BD = CH = EG = IK = JL = MO \text{ is in column } 9$$
$$O = AG = BH = CD = EF = IJ = KL = MN \text{ is in column } 10$$
$$E = AM = BK = CJ = DI = FO = GN = HL \text{ is in column } 11$$
$$F = AN = BL = CI = DJ = EO = GM = HK \text{ is in column } 12$$
$$G = AO = BI = CL = DK = EN = FM = HJ \text{ is in column } 13$$
$$H = AI = BO = CN = DM = EL = FK = GJ \text{ is in column } 14$$
$$I = AH = BG = CF = DE = JO = KN = LM \text{ is in column } 15$$

Main effects and two-factor interaction effects can be estimated using the response table in Figure 3.54. It is recommended that calculations be completed for all columns in Figure 3.54, including columns with no known effects. The estimated effects from columns with no real effects represent observations from probability distributions with average value equal to zero. These "estimates" should be included in a normal plot of estimated effects since they form a reference base (that is, what random variability alone produces) to help in identification of real effects.

Example 4.7

Katy, an environmental engineer, wants to reduce the amount of trihalomethane formation potential (THMFP) in drinking water. A limit of 150 μg per liter has been established for drinking water. Katy has identified twelve factors which could affect the amount of THMFP in water. Previous experiments have indicated that interaction effects among these twelve factors are negligible. Katy has decided to use a sixteen-run fractional factorial experiment for her study. The factors, and their experimental levels, are given in Figure 4.34. The completed response table for the experiment is given in Figure 4.35. The response variable is THMFP concentration measured in $\mu g/\ell$.

The estimated effects are presented graphically in Figure 4.36. Based on this figure, Katy recommended the following levels for factors A, B, C, E, F, I, and L:

$$A_1, B_1, C_1, E_2, F_1, I_1, L_1.$$

The estimated average THMFP at these factor level settings can be calculated by noting what effect each of these factors has on the average response. For example, the average response when A is at level 1 is $\overline{A}_1 = 162.9$. This is $\overline{A}_1 - \overline{y} = 162.9 - 165.8 = -2.9$ units away from the overall average, in a negative direction. If the factors A, B, C, E, F, I, and L are all taken into consideration in this way, the estimated average THMFP is:

$$
\begin{aligned}
\overline{\text{THMFP}} &= \overline{y} + (\overline{A}_1 - \overline{y}) + (\overline{B}_1 - \overline{y}) + (\overline{C}_1 - \overline{y}) + (\overline{E}_2 - \overline{y}) + (\overline{F}_1 - \overline{y}) + (\overline{I}_1 - \overline{y}) + (\overline{L}_1 - \overline{y}) \\
&= 165.8 + (162.9 - 165.8) + (160.6 - 165.8) \\
&\quad + (162.8 - 165.8) + (159.4 - 165.8) \\
&\quad + (161.8 - 165.8) + (161.5 - 165.8) \\
&\quad + (157.9 - 165.8) \\
&= 165.8 - 2.9 - 5.2 - 3.0 - 6.4 - 4.0 - 4.3 - 7.9 \\
&= 132.1 \ \mu g/\ell
\end{aligned}
$$

Factor	Level 1	Level 2
A. Flow rate	5 gpm/ft²	10 gpm/ft²
B. Alum treatment	Adsorption	Sweep
C. Concentration of organic source of color	5 mg/ℓ	25 mg/ℓ
D. Temperature	2° C	20° C
E. Type of filter media	50/50 blend	Grind
F. Addition of non-ionic polymer	0.0 mg/ℓ	0.3 mg/ℓ
G. Addition of cationic polymer	0.0 mg/ℓ	0.3 mg/ℓ
H. Addition of preoxidant K₂MnO₄	0.0 mg/ℓ	2.0 mg/ℓ
I. Addition of powdered activated carbon (PAC)	0.0 mg/ℓ	2.5 mg/ℓ
J. Addition of suspended solids (Kaolin)	0.0 mg/ℓ	10.0 mg/ℓ
K. Type of color source	humic acid	fulvic acid
L. Effect of rapid premixing	alum injection in-line	alum addition to mix tank

FIGURE 4.34 Factors and levels for illustrative example

Random Order Trial Number	Standard Order Trial Number	Response Observed Values y	1 A		2 B		3 C		4 D		5 J		6 K		7 L		8		9		10		11 E		12 F		13 G		14 H		15 I	
			1	2	1	2	1	2	1	2	1	2	1	2	1	2	1	2	1	2	1	2	1	2	1	2	1	2	1	2	1	2
3	1	172	172		172		172		172			172		172		172		172		172		172	172		172		172		172			172
8	2	153	153		153		153			153		153		153	153			153	153		153		153			153		153		153	153	
16	3	155	155		155			155	155			155	155			155	155			155	155			155	155			155		155	155	
2	4	152	152		152			152		152		152	152		152		152		152			152		152		152	152		152			152
13	5	166	166			166	166		166		166			166		166	166		166			166		166		166	166			166	166	
9	6	151	151			151	151			151	151			151	151		151			151	151			151	151			151	151			151
12	7	191	191			191		191	191		191		191			191		191	191		191		191			191		191	191			191
1	8	163	163			163		163		163	163		163		163			163		163		163	163		163		163			163	163	
4	9	146		146	146		146		146		146		146		146			146		146		146		146		146		146	146		146	
5	10	161		161	161		161			161	161		161			161		161	161		161			161	161		161			161		161
7	11	176		176	176			176	176		176			176	176		176			176	176		176			176	176			176		176
10	12	170		170	170			170		170	170			170		170	170		170			170	170		170			170	170		170	
14	13	168		168		168	168		168			168	168		168		168		168			168	168		168			168		168		168
6	14	185		185		185	185			185		185	185			185	185			185	185		185			185	185		185		185	
11	15	154		154		154		154	154			154		154	154			154	154		154			154	154		154		154		154	
15	16	190		190		190		190		190		190		190		190		190		190		190		190		190		190		190		190
TOTAL		2653	1303	1350	1285	1368	1302	1351	1328	1325	1324	1329	1321	1332	1263	1390	1323	1330	1315	1338	1326	1327	1378	1275	1294	1359	1329	1324	1321	1332	1292	1361
NUMBER OF VALUES		16	8	8	8	8	8	8	8	8	8	8	8	8	8	8	8	8	8	8	8	8	8	8	8	8	8	8	8	8	8	8
AVERAGE		165.8	162.9	168.8	160.6	171.0	162.8	168.9	166.0	165.6	165.5	166.1	165.1	166.5	157.9	173.8	165.4	166.3	164.4	167.3	165.8	165.9	172.3	159.4	161.8	169.9	166.1	165.5	165.1	166.5	161.5	170.1
EFFECT			5.9		10.4		6.1		-0.4		0.6		1.4		15.9		0.9		2.9		0.1		-12.9		8.1		-0.6		1.4		8.6	

FIGURE 4.35 Response table for example

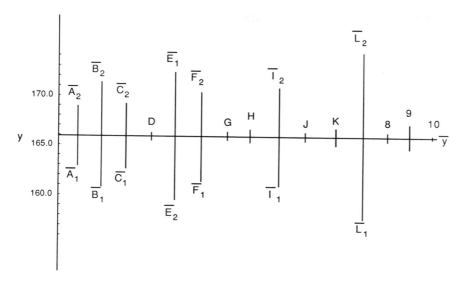

FIGURE 4.36 Graphical display of estimated effects

4.4 Folding Over a Sixteen-run Experimental Design

Folding over eight-run designs, as discussed in section 4.2, enables us to estimate main effects for up to seven factors without confounding these estimates with two-factor interactions. We now explore folding over sixteen-run experiments for much the same reason: they permit us to estimate the main effects for up to fifteen factors without any of these main effects being confounded with two-factor interactions. And this is accomplished with a total of only thirty-two experimental trials.

The procedure for performing a fold-over of a sixteen-run experiment is essentially the same as it was for the eight-run experiment:

1. a. Perform a sixteen-run fractional factorial experiment for up to fifteen factors, following the procedures described in section 4.3.

   b. Analyze the results from the sixteen-run experiment. If the data clearly show which are the important factors and interactions, perform a confirmatory experiment. If confounding causes uncertainty as to which factors and interactions are important, go on to step 2.

2. a. Perform a second sixteen-run experiment based on the design matrix in Figure 4.37. (Figure 4.37 is just Figure 4.33 with 1s and 2s interchanged; that is, Figure 4.37 is a folded-over Figure 4.33. Assign factors to the same columns in Figure 4.37 that they were assigned to in Figure 4.33.)

   b. Complete a response table for these data. Use the form in Figure 3.54, but either reverse the "1" and "2" column headings or write the re-

Standard order	COLUMN NUMBER														
	1	2	3	4	5	6	7	8	9	10	11	12	13	14	15
1	2	2	2	2	1	1	1	1	1	1	2	2	2	2	1
2	2	2	2	1	1	1	2	1	2	2	2	1	1	1	2
3	2	2	1	2	1	2	1	2	1	2	1	2	1	1	2
4	2	2	1	1	1	2	2	2	2	1	1	1	2	2	1
5	2	1	2	2	2	1	1	2	2	1	1	1	2	1	2
6	2	1	2	1	2	1	2	2	1	2	1	2	1	2	1
7	2	1	1	2	2	2	1	1	2	2	2	1	1	2	1
8	2	1	1	1	2	2	2	1	1	1	2	2	2	1	2
9	1	2	2	2	2	2	2	1	1	1	1	1	1	2	2
10	1	2	2	1	2	2	1	1	2	2	1	2	2	1	1
11	1	2	1	2	2	1	2	2	1	2	2	1	2	1	1
12	1	2	1	1	2	1	1	2	2	1	2	2	1	2	2
13	1	1	2	2	1	2	2	2	2	1	2	2	1	1	1
14	1	1	2	1	1	2	1	2	1	2	2	1	2	2	2
15	1	1	1	2	1	1	2	1	2	2	1	2	2	2	2
16	1	1	1	1	1	1	1	1	1	1	1	1	1	1	1

FIGURE 4.37 Design matrix for second half of a folded-over, sixteen-run experiment

sponse values in the shaded spaces instead of on the lines previously used when estimating factor effects. Whichever method you use, be sure that you subtract the average values at level 1 from the average values at level 2 when calculating effects at the bottom of the response table. (See the earlier discussion on this point for eight-run experiments in section 4.2.)

3. Enter the estimated effects and $\bar{y}$ values from the two completed response tables into the second and third columns of a calculation table of the form given in Figure 4.38. Calculate the various estimates as indicated in columns 4 and 5.

4. Interpret the estimates in the last two columns of Figure 4.38 using the information in Figure 4.39. Figure 4.39 assumes there are fifteen factors in the experiment. If there are fewer factors, simply strike all references in Figure 4.39 to the missing factors. For example, if there were only ten factors, they would be labeled A through J. So we would remove all references to factors K, L, M, N, and O, as well as any interaction terms involving these factors. Be sure to do a normal plot of all the estimated effects.

4.5 Blocking Two-level Designs

When an eight-run experiment was folded over in section 4.2, a new column was added to the design matrix which was called *block effect*, or "*Block*," for short (see Figure 4.25). The principle here is that since time had elapsed between performing the two halves of the fold-over design, some uncontrollable and/or unknown factors may have affected the two blocks differently.

Column	Experiment 1 Effect E_1	Experiment 2 Effect E_2	Main Effect $\dfrac{E_1 + E_2}{2}$	Interaction Effect $\dfrac{E_1 - E_2}{2}$
Y				
1				
2				
3				
4				
5				
6				
7				
8				
9				
10				
11				
12				
13				
14				
15				

FIGURE 4.38 Fold-over design effects, calculation table

We may not be particularly interested in estimating the magnitude of such effects, but we don't want them to influence or distort our estimates of other effects. Fortunately, the column for block effect is orthogonal to all the other columns in Figure 4.25, so any block effects can be estimated separately.

The need to block an experiment can occur under a variety of situations

| Row of Figure 4.38 | COLUMN OF FIGURE *4.38* | |
	Main effect	Interaction effect
Y	average	block
1	A	BJ = CK = DL = EM = FN = GO = HI
2	B	AJ = CM = DN = EK = FL = GI = HO
3	C	AK = BM = DO = EJ = FI = GL = HN
4	D	AL = BN = CO = EI = FJ = GK = HM
5	J	AB = CE = DF = GH = IO = KM = LN
6	K	AC = BE = DG = FH = IN = JM = LO
7	L	AD = BF = CG = EH = IM = JN = KO
8	M	AE = BC = DH = FG = IL = JK = NO
9	N	AF = BD = CH = EG = IK = JL = MO
10	O	AG = BH = CD = EF = IJ = KL = MN
11	E	AM = BK = CJ = DI = FO = GN = HL
12	F	AN = BL = CI = DJ = EO = GM = HK
13	G	AO = BI = CL = DK = EN = FM = HJ
14	H	AI = BO = CN = DM = EL = FK = GJ
15	I	AH = BG = CF = DE = JO = KN = LM

FIGURE 4.39 Effects estimated using Figure 4.38

besides fold-over. For example, suppose eight parts must be heat treated, but only four pieces can be treated at a time; or the experiment must be carried out over two shifts; or a given batch can be used in only four trials. What do we do? We assign columns of the design matrix to the block effects. Suppose, for example, an eight-run experiment in four factors had to be split into two four-run blocks. This could be treated as a *five*-factor experiment, with the fifth factor being the block effect. In section 4.1 we found that for five factors in eight runs, factors A, B, C, D, and E should be assigned to columns 1, 2, 3, 4, and 5, respectively. So in order to block a four-factor experiment we could assign the block effect to "factor E," or column 5. Trials with a 1 in column 5 would be assigned to the first block and trials with a 2 in column 5 would be assigned to the second block. From section 4.1, the following alias relationships exist for this five-factor design:

A = BD = CE is in column 1

B = AD is in column 2

C = AE is in column 3

D = AB is in column 4

E = AC is in column 5

BC = DE is in column 6

BE = CD is in column 7.

It is generally safe to assume, however, that block effects do not interact with other factors. That is, there should not be any AE, BE, CE, or DE interaction effects. Then the above alias relationships become:

A = BD in column 1

B = AD in column 2

C is in column 3

D = AB is in column 4

E = AC is in column 5

BC is in column 6

CD is in column 7.

This information can help us intelligently assign factors to the columns of Figure 4.19. For example, the most important variable should probably be labeled "C", since it is the only main effect not confounded with two-factor interactions. If we are particularly interested in the interaction of C with another factor, that factor should be labeled "B" or "D" rather than "A," so that a clean estimate can be obtained for that interaction effect.

Example 4.8

Mary Kay, a metallurgist, wants to improve the strength of a steel product. Four factors are being considered. Eight experimental trials are to be run, but it is possible to run only four per day. Therefore each day will be treated as a separate block. The four experimental factors and their levels are given in Figure 4.40. The completed response table for the experiment is in Figure 4.41. Response is yield strength in psi $\times 10^3$. To clarify the blocking pattern, the blocks to which the runs are assigned are written in the first column of Figure 4.41. The estimated factor effects are plotted in Figure 4.42.

Based on Figures 4.41 and 4.42, factors A, B, and C appear to affect product strength. From section 4.1 we find that, for a five-factor experiment (including block effect) in eight runs, the AB, AC, and BC effects occur in columns 4, 5, and 6 respectively. Apparently there are no interaction effects this time. The block effect is also in column 5: again, no apparent effect. Based on Figures 4.41 and 4.42, A, B, and C should be set at levels 2, 1, and 2, respectively, in order to maximize product strength. At these levels, the estimated average strength is:

$$
\begin{aligned}
\hat{y} &= \bar{y} + (\bar{A}_2 - \bar{y}) + (\bar{B}_1 - \bar{y}) + (\bar{C}_2 - \bar{y}) \\
&= 41.51 + (43.40 - 41.51) + (46.10 - 41.51) + (46.30 - 41.51) \\
&= 41.51 + 1.89 + 4.59 + 4.79 \\
&= 52.78
\end{aligned}
$$

Factor	Level 1	Level 2
A. Composition	0.35%	0.45%
B. Tempering temperature	400° C	500° C
C. Tempering time	1 hour	2 hours
D. Grain size	ASTM no. 4	ASTM no. 6

FIGURE 4.40 Experimental factors and levels

Random Order Trial Number	Standard Order Trial Number	Response Observed Values y	A: Composition		B: Tempering Temperature		C: Tempering Time		D: Grain Size		5 Block		6		7	
			0.35%	0.45%	400°C	500°C	1 Hour	2 Hour	ASTM No.4	ASTM No.6						
Block			1	2	1	2	1	2	1	2	1	2	1	2	1	2
2	1	38.4	38.4		38.4		38.4			38.4		38.4		38.4	38.4	
1	2	49.2	49.2		49.2			49.2		49.2	49.2		49.2			49.2
2	3	30.5	30.5			30.5	30.5		30.5			30.5	30.5			30.5
1	4	40.4	40.4			40.4		40.4	40.4		40.4			40.4	40.4	
1	5	43.7		43.7	43.7		43.7		43.7		43.7			43.7		43.7
2	6	53.1		53.1	53.1			53.1	53.1			53.1	53.1		53.1	
1	7	34.3		34.3	34.3		34.3			34.3	34.3		34.3		34.3	
2	8	42.5		42.5		42.5		42.5		42.5		42.5		42.5		42.5
TOTAL		332.1	158.5	173.6	184.4	147.7	146.9	185.2	167.7	164.4	167.6	164.5	167.1	165.0	166.2	165.9
NUMBER OF VALUES		8	4	4	4	4	4	4	4	4	4	4	4	4	4	4
AVERAGE		41.51	39.63	43.40	46.10	36.93	36.73	46.30	41.93	41.10	41.90	41.13	41.78	41.25	41.55	41.48
EFFECT			3.77		-9.17		9.57		-0.83		-0.77		-0.53		-0.07	

FIGURE 4.41 Response table for example with two blocks

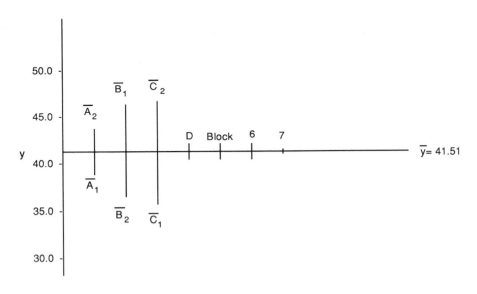

FIGURE 4.42 Estimated factor effects

Instead of subtracting $\bar{y}$ from each "high" factor average in the above equation, we also could have obtained the maximum average strength by adding half of each factor effect to $\bar{y}$, since $\bar{y}$ is halfway between the average responses at the high and low levels of each factor:

$$\hat{y} \; = \; 41.51 + 3.77/2 + 9.17/2 + 9.57/2 \; = \; 52.77.$$

Four blocks

Suppose now that an experiment must be split into four blocks. No problem. Two columns in the design matrix are designated as the block factors and runs are assigned to blocks according to the values in these two columns. For example, suppose we want to run a sixteen-trial experiment in five factors, but have to break the trials into four blocks of four trials each. From section 4.3 we could decide to assign the five design factors, say A, B, C, D, and E, to columns 1, 2, 3, 4, and 11, respectively, of Figure 4.33. Then F and G, columns 12 and 13, could be the block factors. Runs would be assigned to the four blocks according to the scheme in Figure 4.43 (refer to Figure 4.33).

The alias structure for seven factors in sixteen runs is given in section 4.3, and is summarized in column 2 of Figure 4.44. If there are no interactions between block effects and other factors, then by crossing out all interaction terms in column 2 of Figure 4.44 involving F or G, we obtain the reduced set of alias relationships given in column 3 of Figure 4.44. But there is one problem here (read "trap"). The "interaction" between the two block factors F and G is really a main block effect. Referring to Figure 4.33, we find that the FG column (column 8) has 1s in rows 3, 4, 5, 6, 11, 12, 13, and 14, and 2s in the other rows. That is, there are 1s in column 8 of Figure 4.33 for the rows assigned to blocks 2 and 3 in Figure 4.43, and 2s in the rows assigned to blocks 1 and 4. So the FG "interaction" effect is really a main block effect. This means that the AE and BC interaction effects, which by column 2 of Figure 4.44 are confounded with FG, are actually confounded with a *main* block effect. The correct list of alias relationships for the design with five factors in four blocks is given in the last column of Figure 4.44. Does this complicate our lives? Not much. We simply need to be aware of these problems and check *before beginning a blocked experiment* to make sure we have not confounded any block effects, *including interactions between block fac-*

Block	F	G	Trial numbers
1	1	1	1, 8, 10, 15
2	1	2	3, 6, 12, 13
3	2	1	4, 5, 11, 14
4	2	2	2, 7, 9, 16

FIGURE 4.43 Assignment of trials for five-factor experiment to four blocks

Column	7 factors	5 factors in 4 blocks	
		F & G block	F, G, & FG block
5	AB = CE = DF	AB = CE	AB = CE
6	AC = BE = DG	AC = BE	AC = BE
7	AD = BF = CG	AD	AD
8	AE = BC = FG	AE = BC	AE = BC = FG
9	AF = BD = EG	BD	BD
10	AG = CD = EF	CD	CD
15	BG = CF = DE	DE	DE

FIGURE 4.44 Alias structure with five factors and four blocks

tors, with other critical effects. For the case of five factors in four blocks, it is clear that if one factor is the most likely to interact with other factors, it should be labeled "D", since the two-factor interactions involving D are not confounded with any other main or two-factor interaction terms.

If we wanted to break an experiment into eight blocks, we would need to use three columns of the design matrix for blocking factors. If we call these three factors E, F, and G, then the columns for EF, EG, FG, *and* EFG would also be confounded with the blocks. Again, careful checking is called for before starting the experiment.

Example 4.9

An engineer is interested in the effects of cutting speed (A), cutting angle (B), tool hardness (C), and cutting tool angle (D) on the life of a tool. Two levels are selected for each factor, and sixteen runs are planned. The response variable will be tool life, measured in hours. Because of the anticipated length of time required for each run, it was decided to use four different stations (that is, four blocks) simultaneously for the machining operation.

Based on section 4.3, it was decided to assign the four design factors, A, B, C, and D, to columns 1, 2, 3, and 4, respectively, of Figure 4.33. Factors E and F, columns 11 and 12, were made the block factors. Runs were then assigned to blocks according to the scheme in Figure 4.45. The confounding

Block	E	F	Trial numbers
1	1	1	1, 8, 12, 13
2	1	2	2, 7, 11, 14
3	2	1	3, 6, 10, 15
4	2	2	4, 5, 9, 16

FIGURE 4.45 Assignment of trials for four-factor experiment to four blocks

Column	With block interactions	Without block interactions
5	AB = CE = DF	AB
6	AC = BE	AC
7	AD = BF	AD
8	BC = AE	BC
9	BD = AF	BD
10	CD = EF	CD = EF
15	CF = DE	—

FIGURE 4.46 Alias structure with four factors and four blocks

relationships for main and two-factor interactions for the experiment are given in Figure 4.46. Column 2 gives the alias structure allowing for block interaction effects, and column 3 gives the alias structure assuming no block interaction effects. Remember that "EF" is a main block effect, not an interaction.

The completed response table for the experiment is given in Figure 4.47. Based on the estimated effects calculated there, tool life can be maximized by setting factor A at level 2 and factor C at level 1. The estimated average life would then be

$$\hat{y} = \bar{y} + (\bar{A}_2 - \bar{y}) + (\bar{C}_1 - \bar{y})$$
$$= 25.1 + (31.4 - 25.1) + (31.9 - 25.1)$$
$$= 38.1$$

For this example, the block effects (columns 10, 11, and 12 of Figure 4.47) are negligible. However, if there were significant blocking effects, the easiest way to analyze them is to calculate the average value of the response for each block, and then compare these four averages.

4.6 Other Useful Two-level Designs

No experimental designs are more valuable to the engineer than those covered in this chapter. But we have certainly not exhausted the possibilities. In Chapter 7 experimental designs will be presented in which factors are set at three levels. Other designs allow any number of levels. Using more than two levels per factor allows testing for nonlinearity in the relationship between a factor and the response variable. But dealing with interaction effects then becomes more complicated. We looked at folding over an eight-run experiment. We can fold over any two-level orthogonal design to improve design resolution. We looked at four-, eight-, and sixteen-run experiments. We can easily extend this to thirty-two, sixty-four, or more trials. There is a whole class of two-level designs called the Placket and Burman designs, which allow esti-

FIGURE 4.47 Response table for experiment with four factors and four blocks

Column index / effect labels: 1 = A, 2 = B, 3 = C, 4 = D, 5 = AB, 6 = AC, 7 = AD, 8 = BC, 9 = BD, 10 = CD + Block, 11 = Block, 12 = Block, 13, 14, 15

Std. Order Trial No.	Block	Response Y	A·1	A·2	B·1	B·2	C·1	C·2	D·1	D·2	AB·1	AB·2	AC·1	AC·2	AD·1	AD·2	BC·1	BC·2	BD·1	BD·2	CD+Block·1	CD+Block·2	Block(11)·1	Block(11)·2	Block(12)·1	Block(12)·2	13·1	13·2	14·1	14·2	15·1	15·2
1	1	23	23		23		23		23			23		23		23		23		23		23	23		23		23			23			23
2	2	25	25		25		25			25		25		25	25			25	25		25		25			25		25		25	25	
3	3	10	10		10			10	10			10	10			10	10			10	10			10	10			10		10	10	
4	4	12	12		12			12		12		12	12		12		12		12			12		12		12	12		12			12
5	4	24	24			24	24		24		24			24		24	24		24			24		24		24	24			24	24	
6	3	30	30			30	30			30	30			30	30		30			30	30			30	30			30	30			30
7	2	13	13			13		13	13		13		13			13		13	13		13		13			13		13	13			13
8	1	14	14			14		14		14	14		14		14			14		14		14	14		14		14			14	14	
9	4	37		37	37		37		37		37		37		37			37		37		37		37		37		37	37		37	
10	3	38		38	38		38			38	38		38			38		38	38		38			38	38		38			38		38
11	2	24		24	24			24	24		24			24	24		24			24	24		24			24	24			24		24
12	1	25		25	25			25		25	25			25		25	25		25			25	25		25			25	25		25	
13	1	38		38		38	38		38			38	38		38		38		38			38	38		38			38		38		38
14	2	40		40		40	40			40		40	40			40	40			40	40			40		40	40		40		40	
15	3	22		22		22		22	22			22		22	22			22	22		22			22	22		22		22		22	
16	4	27		27		27		27		27		27		27		27		27		27		27		27		27		27		27		27
TOTAL		**402**	151	251	194	208	255	147	191	211	205	197	202	200	202	200	203	199	197	205	202	200	202	200	200	202	197	205	202	200	197	205
NUMBER OF VALUES		16	8	8	8	8	8	8	8	8	8	8	8	8	8	8	8	8	8	8	8	8	8	8	8	8	8	8	8	8	8	8
AVERAGE		25.1	18.9	31.4	24.3	26.0	31.9	18.4	23.9	26.4	25.6	24.6	25.3	25.0	25.3	25.0	25.4	24.9	24.6	25.6	25.3	25.0	25.3	25.0	25.0	25.3	24.6	25.6	25.3	25.0	24.6	25.6
EFFECT			12.5		1.7		-13.5		2.5		-1.0		-0.3		-0.3		-0.5		1.0		-0.3		-0.3		0.3		1.0		-0.3		1.0	

mation of k-1 main factor effects in k trials when k is a multiple of 4 and does not exceed 100. The alias structures of the Placket and Burman designs can be difficult to obtain, but when folded over they produce designs of resolution IV. The reader is referred to section 9.1, and to Placket and Burman (1946), Box, Hunter, and Hunter (1978), or Lochner and Matar (1988) for more on these designs.

Evaluating Variability

5

5.1 Why Analyze Variability?

In Chapters 3 and 4 we saw how to use experimental designs to increase yields, improve welds, decrease travel times, improve surface finish. All the examples and discussion focused on improving performance *on the average*. That is, maximize the average yield or decrease the average travel time. But there is more to quality than averages. If average plating thickness is increased by ten percent, but the plating is not as even as before, and has more thin spots, then thicker is not necessarily better (see Figure 5.1). Or if a new

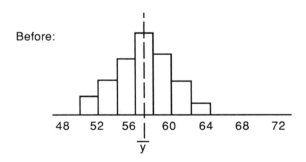

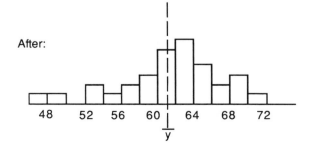

FIGURE 5.1 Increase in plating thickness average and variability

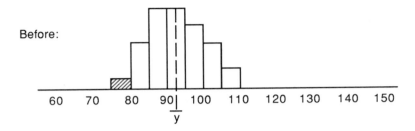

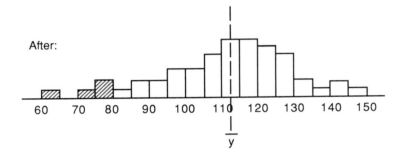

FIGURE 5.2 Increase in average strength and in proportion below 80 lbs

treatment increases average weld strength by twenty percent but also causes a larger percentage to break at 80 pounds of force, the new treatment is not an improvement (see Figure 5.2).

In manufactured products, amount of variability is critical for fit and finish. Ford Motor Company found that when two plants produced automobile transmissions to the same specifications, the plant with minimal variability in product characteristics had fewer customer complaints, *even though both plants were manufacturing within the same specifications*. Here, as in many other examples which could be cited, reduced variability meant reduced costs to the manufacturer and to the user. A commitment to never-ending improvement must mean a commitment to decreasing variability in all processes and products.

Observed response values are usually different from trial to trial in an experiment. Some of the variability is the result of changing the levels of the experimental factors. Some of the variability is due to random variation in the process. Even under ideal conditions—same production line, same operator, same amounts of the same ingredients at the same temperature for the same amount of time—the "same" experiment will produce a slightly different result each time it is performed. Randomness is an inherent part of any process. If we are to improve product and process quality and performance, we need to understand randomness, measure it, and seek ways to reduce its effects. By reducing process variability we can better control processes and can reduce the costs associated with development, manufacture, and use.

Measures of process and product variability are evaluated in much the same way as were measures of process average in Chapters 3 and 4. This chapter will be fairly easy in that no new experimental designs will be introduced, and we consider only the case of *reducing* variability (smaller-the-better).

5.2 Measures of Variability

In quality control and improvement, two measures of variability of sample data are commonly used:

- R, the sample range, and
- s, the sample standard deviation.

The *sample range* is the arithmetic difference between the largest and smallest values in the sample. For example, consider the following sample of six measurements:

172, 164, 167, 179, 172 and 175.

The sample range is equal to

$$
\begin{aligned}
R &= \text{largest} - \text{smallest} \\
&= 179 - 164 \\
&= 15.
\end{aligned}
$$

Although the sample range is easy to calculate and to understand, it has some serious drawbacks as a general measure of variability:

1. It uses only the largest and smallest numbers in the sample. This makes it somewhat unstable since a single outlier (observed value far removed from most of the other sample values) can result in an artificially high value for the sample range.
2. Since it is calculated from only two of the sample values (the largest and the smallest), it does not reflect the variability of the other measurements.
3. The likelihood of obtaining a large value for the sample range increases as the size of the sample increases. So ranges from samples of different sizes cannot be directly compared.

The *sample standard deviation* is the generally accepted measure of variability in statistical data analysis and experimental design. This statistic is somewhat more difficult to calculate than the sample range, but it has desirable properties which make its use worth the added effort. And if a computer is used in data analysis, calculation time for either measure of variability becomes insignificant. There are five steps in calculating the sample standard deviation:

1. Calculate the sample mean, $\bar{y}$.
2. Subtract $\bar{y}$ from each measurement in the sample.

3. Square the differences obtained in step 2.
4. Add the squared differences obtained in step 3 and divide the sum by the sample size minus one. This statistic is called the sample variance and is denoted by s^2.
5. Obtain the square root of s^2. This is the sample standard deviation and is denoted by s.

The calculation of the sample variance can be written in a single formula using the "Σ" notation for adding up a sequence of values:

$$s^2 = \Sigma(y-\bar{y})^2/(n-1).$$

Then:

$$s = \sqrt{s^2}$$

Example

Calculate the standard deviation for the following six measurements:

172, 164, 167, 179, 172, 175

The average or mean of these six measurements is:

$$\bar{y} = \frac{172+164+167+179+172+175}{6}$$
$$= \frac{1029}{6}$$
$$= 171.5$$

Steps 2 and 3 are performed in Figure 5.3. Note in Figure 5.3 that the sum of the differences, $\Sigma(y-\bar{y})$, is equal to zero. This is always true since $\Sigma(y-\bar{y})=\Sigma y-\Sigma\bar{y}=\Sigma y-n\bar{y}$, and $\bar{y}=\Sigma y/n$.
The sample variance is equal to

Observed value, y	Sample mean, $\bar{y}$	Difference $y-\bar{y}$	Squared $(y-\bar{y})^2$
172	171.5	0.5	0.25
164	171.5	-7.5	56.25
167	171.5	-4.5	20.25
179	171.5	7.5	56.25
172	171.5	0.5	0.25
175	171.5	3.5	12.25
Sum:		0.0	145.50

FIGURE 5.3 Calculation of sum of squared differences

$$s^2 = \frac{145.50}{5}$$
$$= 29.1$$

and the sample standard deviation is:

$$s = \sqrt{29.1}$$
$$= 5.39.$$

5.3 The Normal Distribution

The sample standard deviation may look a little complicated. Why can't we use a simpler measure of variability instead? One reason the standard deviation is so important is because of the role it plays in what is called the *normal distribution*. As was pointed out at the beginning of this chapter, random variability is a part of any process. If repeated measurements are taken of some process characteristic over time, different values are observed, even for stable processes. If a set of such process measurements were collected, tabulated, and plotted as a bar chart or *frequency diagram*, they would form some sort of a pattern. If another set of measurements were taken from the same process and also tabulated, this second data set would display a different pattern or *sample distribution*. But if both data sets were large, say one or two hundred measurements each, the two distributions would be similar. Although we cannot predict in advance the exact value for any particular measurement, we can predict overall patterns. If you toss a coin once, you have no idea whether it will land heads or tails. But in 200 tosses, we can reasonably expect to get between 80 and 120 heads. Actuaries don't know who will die in the next year, but they can make surprisingly good predictions of the *number* who will die in any particular population or subpopulation. Figure 5.4 gives the amount of sodium, in mg/oz, from twenty-four tests of a food product. The data are presented in a frequency diagram in Figure 5.5. A second set of twenty-four measurements is given in Figure 5.6, with a corresponding frequency diagram in Figure 5.7. We can see a similarity between Figures 5.5 and 5.7. Something else we may notice is that the two frequency diagrams have roughly a bell-shaped appearance. If we include the data from both samples in the same frequency diagram (Figure 5.8), the bell shape is even more clear.

Not all data sets produce bell-shaped frequency diagrams. But a surprising number of samples of process measurements do. An important result from mathematical statistics, called the *Central Limit Theorem*, says that if you repeatedly take large random samples from a stable process, calculate the sample mean for each sample, and display these averages in a frequency diagram, the diagram will be roughly bell-shaped. (There is more to the Central Limit Theorem than this, but this is enough for our needs right now.) With many stable processes, observed measurement variability is due to many sources of variation, each contributing a small amount to the overall variabil-

192	191	188	187	188	189	186	188
186	189	194	190	184	188	182	185
189	191	190	189	187	187	188	188

FIGURE 5.4 Amount of sodium in sample items (mg/oz)

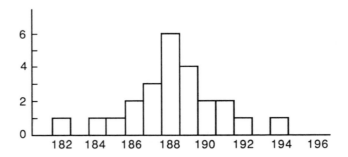

FIGURE 5.5 Frequency diagram of data in Figure 5.4

ity. The "adding up" of these individual contributions sometimes acts like the summing of values when calculating a sample mean, and so the Central Limit "effect" may appear with the individual measurements. The Central Limit Theorem provides the formula for the bell-shaped curve, which is also called the *normal probability distribution*. We don't actually need the for-

188	191	189	187	185	190	189	187
193	188	189	184	194	189	189	196
188	187	190	185	188	193	190	186

FIGURE 5.6 Second sample of sodium measurements

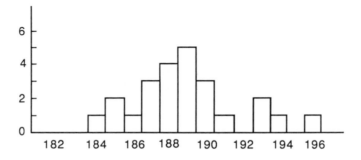

FIGURE 5.7 Frequency diagram of data in Figure 5.6

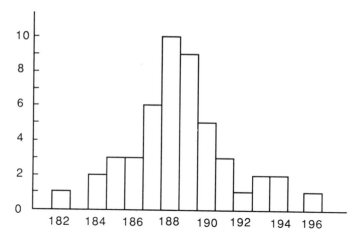

FIGURE 5.8 Frequency diagram of data in Figure 5.4 and 5.6

mula or become involved in calculations with it. But we should understand that:

- The normal or bell-shaped curve represents the limiting form for the distribution of sample data from a process if we took thousands of measurements and either the measurements themselves were normally distributed, or else the "measurements" were averages of large samples of independently obtained values. To actually "fit" a normal curve to a sample frequency diagram the scale for the vertical axis would have to be the proportion of the sample instead of actual frequency (that is, divide frequency by sample size to get the new scale).
- Although the Central Limit Theorem technically applies for only very large samples, for many processes the averages of rather small samples, and even estimated effects as calculated in Chapters 3 and 4, will tend to follow a normal distribution. This is our justification for plotting factor effects on normal probability paper in section 3.6.

The shape of a normal curve is determined by two numbers—the mean and the standard deviation. We sometimes speak of a stable process as having fixed but unknown values for these two *parameters*. We denote the average by the Greek letter μ, pronounced "mu," and the standard deviation by the Greek letter σ, pronounced "sigma." We can think of them as being calculated in the same way as the sample mean and standard deviation, $\bar{y}$ and s, but using all the possible measurements which could be taken from the process. The normal distribution is symmetric and centered at μ. σ measures the dispersion of the process. Larger σ means more spread. (See Figure 5.9.)

Since we cannot usually obtain all possible process measurements, and so cannot actually calculate μ and σ, we take a sample of measurements from the process instead and estimate the process *parameters* μ and σ using the

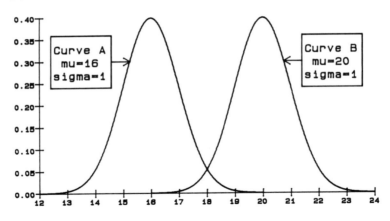

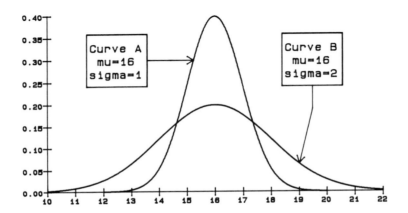

FIGURE 5.9 Normal probability curves

sample *statistics* $\bar{y}$ and s. Because of random variability, $\bar{y}$ and s will vary from one sample to the next, even if the process is stable.

You may be wondering where all this is going: why was this business of μ and σ brought up at all? Since the normal distribution is completely described by its mean and standard deviation, we can predict what proportion of measurements will fall within any particular interval once μ and σ are known. (Although we don't know the true values of μ and σ, we can estimate them by $\bar{y}$ and s.) About 68 percent of the process measurements will fall between $\mu - \sigma$ and $\mu + \sigma$. Similarly, about 95 percent of the process measurements will fall between $\mu - 2\sigma$ and $\mu + 2\sigma$. Over 99 percent will fall between $\mu - 3\sigma$ and $\mu + 3\sigma$. This is illustrated in Figure 5.10.

Example

The sample data in Figure 5.4 have a sample mean of $\bar{y} = 188.2$ and a sample standard deviation $s = 2.6$. (The reader may want to verify these values using

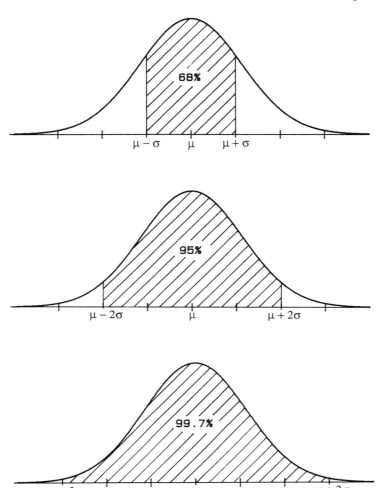

FIGURE 5.10 Percentage of observations in intervals about the mean of a normal distribution, expressed in terms of the standard deviation

the formulas presented in section 5.2.) If the data are independent measurements from a stable bell-shaped population, we can estimate that:

- About 68 percent of the process measurements lie between $188.2 - 2.6 = 185.6$ and $188.2 + 2.6 = 190.8$.
- About 95 percent of the process measurements lie between $188.2 - (2 \times 2.6) = 188.2 - 5.2 = 183.0$ and $188.2 + (2 \times 2.6) = 188.2 + 5.2 = 193.4$.
- Over 99 percent of the process measurements lie between $188.2 - (3 \times 2.6) = 188.2 - 7.8 = 180.4$ and $188.2 + (3 \times 2.6) = 188.2 + 7.8 = 196.0$.

This can be illustrated using Figure 5.10 if we replace μ and σ by their sample estimates, $\bar{y}$ and s.

5.4 Using Two-level Designs to Minimize Variability

In Chapters 3 and 4 we saw how to use two-level experimental designs to measure the effects factors have on average values of response variables. Now we will use these same experimental designs to measure effects of factors on *variability* of response variables. The approach is very similar to what was done in Chapters 3 and 4. Factors can affect a response variable in two basic ways: by changing the average value or by changing the amount of variability. We already know how to analyze for changes in average value. To analyze for changes in variability it is necessary to estimate the amount of variability at each "point" in the experimental design. This will necessitate obtaining repeat observations for each combination of factors used in the experiment. That is, we will need to replicate experiments. And, in order to avoid confounding block effects with variation effects, the order in which the experimental trials are performed should be randomized as a single group. When dealing with replicated experiments in Chapter 3, we calculated an average response at each experimental point and worked with these averages as if they were single observations. We will do much the same thing here, but instead of calculating averages we will calculate standard deviations at each point. The approach to analyzing the factor effects on response variability can be summarized by the following seven steps:

1. Select an appropriate experimental design. Use the same basic criteria that were used in Chapters 3 and 4.

2. Determine the number of replications to be used. Don't be shy here. Although a sample standard deviation can be calculated using only two measurement values, five to ten replicates provide much more reliable estimates of response variability. Cost considerations frequently determine the number of trials which can be run. If only two or three replicates can be afforded, the experimenter must exercise caution when interpreting the results.

3. Randomize the order in which the trials will be performed. Some people do all the replications for a particular setting of factor levels in sequence, since this will simplify the data gathering process. This is often a mistake, since data gathered over a short period of time may not reflect true process variation, and artificially low sample standard deviations may result.

4. Perform the experiment; record the sample values.

5. Group together the experimental run data for each factor level combination used in the experiment. Calculate a sample standard deviation separately for each combination of factor levels.

6. Calculate the logarithms of the standard deviations obtained in step 5. (The rationale for this step is explained below.) Record these logarithms in the appropriate locations of the "Response observed values" column of a response table.

7. Complete and analyze the response table in the same manner as was done in Chapters 3 and 4, but use the logarithms of the standard deviations as

the raw response values. Keep in mind that for variability the rule is always smaller-the-better.

Now to the matter of why logarithms of the sample standard deviations are used instead of the standard deviations themselves. In section 5.3 we mentioned that the Central Limit effect often results in response variables which are roughly normally distributed. When this happens, the sample averages and estimated effects are also approximately normally distributed, *but the sample standard deviations are not*. However, if we transform the sample standard deviations by taking their logarithms (natural logs, base 10, or whatever), the logs of the standard deviations will be much closer to being normally distributed. Hence, they can be plotted using normal probability paper and generally treated the same as sample means.

Example 5.1

In Section 4.1 an example was given which involved maximizing bond strength when mounting an integrated circuit (I.C.) on a metallized glass substrate. An eight-run experiment in four factors was performed. The factors and their levels are listed in Figure 4.5. That figure is reproduced as Figure 5.11. Two of the factors (cure time and I.C. post coating) were found to affect average bond strength. But when bonding I.C.'s, it is not enough to have the bond strong *on the average*. *Every* I.C. must be securely fastened. So increasing average bond strength at the cost of increased variability in those bonds may be counterproductive. We will continue the illustrative example from section 4.1 by exploring the effects of the experimental factors on bond strength variability.

The combinations of factor levels to be used are given in Figure 5.12. They are the same as those used earlier (Figure 4.6), but now the experiment will be replicated five times. This means there will be (8 level combinations) × (5 replications) = 40 experimental trials. These forty trials should be performed in randomized order. The data from the experiment, grouped by factor level combinations, are given in Figure 5.13. The sample average, standard deviation, and base 10 logarithm of the standard deviation for each set of five replications are given in the last three columns of Figure 5.13.
The response table in Figure 5.14 is based on the sample means from Figure 5.13. The response table in Figure 5.15 uses the logarithms (base 10) of the standard deviations from Figure 5.13 as the observed response values. The estimated factor effects on average values which are calculated at the bottom

Factor	Low level	High level
A. Adhesive type	D2A	H-1-E
B. Conductor material	copper	nickel
C. Cure time (at 90° C)	90 min.	120 min.
D. I.C. post coating	tin	silver

FIGURE 5.11 Factor levels for example experiment

Standard order number	Adhesive type	Conductor material	Cure time	I.C. post coating
1	D2A	copper	90	tin
2	D2A	copper	120	silver
3	D2A	nickel	90	silver
4	D2A	nickel	120	tin
5	H-1-E	copper	90	silver
6	H-1-E	copper	120	tin
7	H-1-E	nickel	90	tin
8	H-1-E	nickel	120	silver

FIGURE 5.12 Experimental design for example

Standard order number	Observed response values	Sample mean, $\bar{y}$	Sample standard deviation, s	Logarithm of standard deviation, $\log s$
1	73.0, 73.2, 72.8, 72.2, 76.2	73.48	1.57	0.196
2	87.7, 86.4, 86.9, 87.9, 86.4	87.06	0.71	−0.149
3	80.5, 81.4, 82.6, 81.3, 82.1	81.58	0.80	−0.097
4	79.8, 77.8, 81.3, 79.8, 78.2	79.38	1.41	0.149
5	85.2, 85.0, 80.4, 85.2, 83.6	83.88	2.06	0.314
6	78.0, 75.5, 83.1, 81.2, 79.9	79.54	2.93	0.467
7	78.4, 72.8, 80.5, 78.4, 67.9	75.60	5.17	0.713
8	90.2, 87.4, 92.9, 90.0, 91.1	90.32	1.99	0.299

FIGURE 5.13 Data from replicated example

of Figure 5.14 confirm the results from Figure 4.7, which were based on an unreplicated experiment. In particular, factors C and D should both be set at level 2 in order to maximize average bond strength. The estimated log s effects from Figure 5.15 are plotted in Figure 5.16. Factors A (adhesive type) and D (I.C. post coating) affect bond strength variability. In order to minimize variability, adhesive type 1 (D2A) and I.C. post coating 2 (silver) should be used. So there are no conflicts here between levels recommended to maximize average strength and levels recommended to minimize variability. Note also that although factor C does not appear significantly to affect variability, the data suggest (Figure 5.15) that level 2 may result in somewhat lower variability.

5.5 Signal-to-Noise Ratio

Dr. Taguchi proposed a class of statistics called *signal-to-noise ratios*. These S/N ratios are meant to be used as measures of the effect of noise factors on performance characteristics. S/N ratios take into account both amount of vari-

Response table for average response (Figure 5.14):

#	y' Bond Strength Average	A: Adhesive Type — DZA (1)	A: H-I-E (2)	B: Conductor material — Cu (1)	B: Ni (2)	C: Cure Time — 90 min (1)	C: 120 min (2)	AB: (1)	AB: (2)	AC: (1)	AC: (2)	BC: (1)	BC: (2)	D: I.C. Post Coating — tin (1)	D: silver (2)
1	73.48	73.48		73.48		73.48			73.48		73.48		73.48	73.48	
2	87.06	87.06		87.06			87.06	87.06			87.06	87.06			87.06
3	81.58	81.58			81.58	81.58		81.58			81.58	81.58			81.58
4	79.38	79.38			79.38		79.38	79.38		79.38			79.38	79.38	
5	83.88		83.88	83.88		83.88		83.88		83.88			83.88		83.88
6	79.54		79.54	79.54			79.54	79.54			79.54	79.54		79.54	
7	75.60		75.60		75.60	75.60			75.60	75.60		75.60		75.60	
8	90.32		90.32		90.32		90.32		90.32		90.32		90.32		90.32
TOTAL	650.84	321.50	329.34	323.96	326.88	314.54	336.30	324.38	326.46	325.92	324.92	323.78	327.06	308.00	342.84
NUMBER OF VALUES	8	4	4	4	4	4	4	4	4	4	4	4	4	4	4
AVERAGE	81.36	80.38	82.34	80.99	81.72	78.64	84.08	81.10	81.62	81.48	81.23	80.95	81.77	77.00	85.71
EFFECT		1.96		0.73		5.44		0.52		−0.25		0.82		8.71	

FIGURE 5.14 Response table for average response

Response table for log s (Figure 5.15):

#	y' Bond Strength Log s	A: Adhesive Type — DZA (1)	A: H-I-E (2)	B: Conductor material — Cu (1)	B: Ni (2)	C: Cure Time — 90 min (1)	C: 120 min (2)	AB: (1)	AB: (2)	AC: (1)	AC: (2)	BC: (1)	BC: (2)	D: I.C. Post Coating — tin (1)	D: silver (2)
1	0.196	.196		.196		.196			.196		.196		.196	.196	
2	−0.149	−.149		−.149			−.149	−.149			−.149	−.149			−.149
3	−0.097	−.097			−.097	−.097		−.097			−.097	−.097			−.097
4	0.149	.149			.149		.149	.149		.149			.149	.149	
5	0.314		.314	.314		.314		.314		.314			.314		.314
6	0.467		.467	.467			.467	.467			.467	.467		.467	
7	0.713		.713		.713	.713			.713	.713		.713		.713	
8	0.299		.299		.299		.299		.299		.299		.299		.299
TOTAL	1.892	.099	1.793	.828	1.064	1.126	.766	.833	1.059	1.027	.865	.934	.958	1.525	.367
NUMBER OF VALUES	8	4	4	4	4	4	4	4	4	4	4	4	4	4	4
AVERAGE	0.237	0.025	0.448	0.207	0.266	0.282	0.192	0.208	0.265	0.257	0.216	0.234	0.240	0.381	0.092
EFFECT		0.423		0.059		−0.090		0.057		−0.041		0.006		−0.289	

FIGURE 5.15 Response table for log s

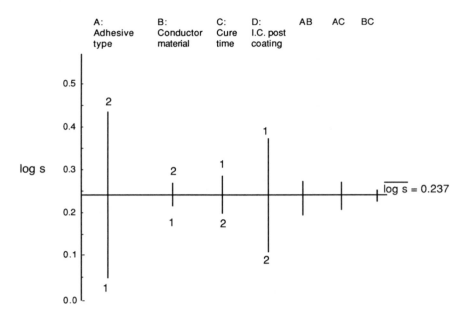

FIGURE 5.16 Plot of factor effects on log(s)

ability in the response data and closeness of the average response to target. Taguchi proposed over seventy such statistics (see Barker, 1986, and Phadke, 1989). Three of them have application to a wide range of situations and will be considered here. They are the S/N ratios for: smaller-is-better, larger-is-better and nominal-is-best (or some use "target-value-is-best").

Smaller-is-better

The "target" value for the response is zero. Recall the formula for the sample variance:

$$s^2 = \Sigma(y - \overline{y})^2/(n-1)$$

If $\overline{y}$ is replaced by its target value, then the formula for s^2 becomes:

$$s^2 = \Sigma(y - 0)^2/(n-1)$$

As the average response value decreases we can expect this function of observed response values to decrease. Taguchi used the following function of the numerator in the above expression as his S/N ratio for smaller-the-better situations:

$$\text{S/N ratio} = -10 \log (\Sigma y^2/n)$$

The goal of an experiment for *smaller-is-better* situations is to minimize Σy^2. This is accomplished by maximizing $-10 \log (\Sigma y^2/n)$.

Larger-is-better

Now the goal is to *maximize* the response. (The response could be yield, for example.) But to maximize y is to minimize 1/y. So Taguchi proposed the following S/N ratio for the larger is better case:

$$\text{S/N ratio} = -10 \log \left(\Sigma(1/y^2)/n\right)$$

Again, the object is to maximize the S/N ratio.

Nominal-is-best

For this situation Taguchi recommends the following form for the S/N ratio:

$$\text{S/N ratio} = 10 \log(\bar{y}^2/s^2)$$

As the variability of the response gets smaller, relative to the average response, this S/N ratio will increase.

There has been a great deal of debate regarding the appropriateness of using these S/N ratios as measures of process or product quality. (See Box, 1988, and Pignatiello and Ramberg, 1985.) Many Western statisticians feel that it is better to use two separate performance statistics—one to measure average response and a second to measure response variability. Sometimes response variability changes as a function of response average. For other response variables, average and variability appear to be independent. By looking at averages and standard deviations separately, rather than depending on signal-to-noise ratios as the sole performance criterion, the engineer is able to make more informed choices. Two examples of use of S/N ratios are given in this chapter. One is below and the other is in the next section.

Example 5.2

We return once more to the example concerned with maximizing bond strength when mounting integrated circuits. The example was first introduced in section 4.1, and was developed further in section 5.4. Four factors were identified as possibly affecting bond strength. They are listed in Figure 5.11. Analysis of the data in section 5.4 indicated that strength could be maximized and variability minimized by the following assignment of factor levels:

- factor A, adhesive type, at level 1,
- factor C, cure time, at level 2, and
- factor D, post coating, at level 2.

The sample response values in Figure 5.13 will now be analyzed using S/N ratios. The S/N ratio formula for the larger-the-better case will be used since the goal of the experiment is to maximize bond strength. The sample S/N ratios are given in Figure 5.17.

The S/N ratios from Figure 5.17 have been entered as response values

Standard order number	$\Sigma(1/y^2)/n$	S/N ratio = $-10 \log (\Sigma(1/y^2)/n)$
1	0.0001854	37.32
2	0.0001320	38.79
3	0.0001503	38.23
4	0.0001588	37.99
5	0.0001423	38.47
6	0.0001586	38.00
7	0.0001771	37.52
8	0.0001227	39.11

FIGURE 5.17 S/N ratios for example data

into the response table in Figure 5.18. From the effects calculated at the bottom of Figure 5.18 it appears that in order to maximize the S/N ratio, the following assignments should be made:

- factor A, adhesive type, at level 2,
- factor C, cure time, at level 2, and
- factor D, post coating, at level 2.

Random Order Trial Number	Standard Order Trial Number	Response Observed Values y · S/N Ratio	A: Adhesive Type D2A 1	H-I-E 2	B: Conductor Material Cu 1	Ni 2	C: Cure Time 90min 1	120min 2	AB: 1	2	AC: 1	2	BC: 1	2	D: Post Coating tin 1	Silver 2
	1	37.32	37.32		37.32		37.32			37.32		37.32		37.32	37.32	
	2	38.79	38.79		38.79			38.79	38.79	38.79		38.79				38.79
	3	38.23	38.23			38.23	38.23		38.23			38.23	38.23			38.23
	4	37.99	37.99			37.99		37.99	37.99		37.99			37.99	37.99	
	5	38.47		38.47	38.47		38.47		38.47		38.47			38.47		38.47
	6	38.00		38.00	38.00			38.00	38.00			38.00	38.00		38.00	
	7	37.52		37.52		37.52	37.52			37.52	37.52		37.52		37.52	
	8	39.11		39.11		39.11		39.11		39.11		39.11		39.11		39.11
TOTAL		305.43	152.33	153.10	152.58	152.85	151.54	153.89	152.69	152.74	152.77	152.66	152.54	152.89	150.83	154.60
NUMBER OF VALUES		8	4	4	4	4	4	4	4	4	4	4	4	4	4	4
AVERAGE		38.18	38.08	38.28	38.15	38.21	37.89	38.47	38.17	38.19	38.19	38.17	38.14	38.22	37.71	38.65
EFFECT			0.20		0.06		0.58		0.02		−0.02		0.08		0.94	

FIGURE 5.18 Response table for S/N ratios

This is the same recommendation for C and D, but the opposite recommendation for A, when compared with our earlier results. But the S/N ratio for factor A is not exceptionally large, and so is not conclusive. Figure 5.16, however, shows strong evidence that variability is reduced when factor A is set at level 1.

5.6 Minimizing Variability and Optimizing Averages

In this section an example involving fifteen factors will be used to illustrate application of the concepts presented in this chapter. The data are from a real case study done by Jim Quinlan of Flex Products, Inc. The study was originally presented in 1985 at the Third Supplier Symposium on Taguchi Methods. The American Supplier Institute selected this paper as the best case study of 1985.

Example 5.3

The product and quality characteristic of concern were described as follows in Quinlan's paper:

> The product under test in this experiment was extruded thermoplastic speedometer casing. . . . This product is used to cover the mechanical speedometer cable on automobiles. The product consists of an extruded polypropylene inner liner, a layer of braided wire, and a coextruded casing. This product has been produced for over fifteen years. Prior to manufacture by Flex Products, the casing under test had been produced by a division of General Motors Corporation. That division had conducted much one factor at a time experimentation with high costs and disappointing results.
>
> The quality characteristic of concern is the post extrusion shrinkage of the casing. Excessive shrinkage can cause noise in the assembly, which has been one of the larger problems with mechanical speedometer cable assemblies. The post extrusion shrinkage is approximated with a two hour heat soak test. . . . The percent shrinkage is obtained by measuring a length of casing that has been properly conditioned, placing that casing in a two hour heat soak in an air circulating oven, reconditioning the sample, and measuring the length. The post test length is then subtracted from the original length, divided by the original length, and then multiplied by 100 to obtain a percent result. [The response variable is percent shrinkage.] The approximate length of samples is 600 mm.

The group at Flex Products working on this problem obtained input from customers, production, quality control, product engineers, and process engineers in compiling their list of factors most likely to affect casing shrinkage. The fifteen most likely candidates are listed in Figure 5.19, along with the

Factor	Level 1	Level 2	Column
A. Liner O.D.	Changed	Existing	1
B. Liner die	Changed	Existing	2
C. Liner line speed	80% of existing	Existing	3
D. Liner tension	More	Existing	4
E. Wire diameter	Existing	Smaller	11
F. Coating die type	Changed	Existing	12
G. Screen pack	Denser	Existing	13
H. Cooling method	Changed	Existing	14
I. Line speed	70% of existing	Existing	15
J. Liner material	Changed	Existing	5
K. Wire braid type	Changed	Existing	6
L. Liner temperature	Preheated	Ambient	7
M. Braiding tension	Changed	Existing	8
N. Coating material	Changed	Existing	9
O. Melt temperature	Cooler	Existing	10

FIGURE 5.19 Factors and experimental levels

levels to be used in the experiment. (Numerical values for levels were not reported in the paper. Also, the ordering of factors in Figure 5.19 differs from Quinlan's paper, and his factor level 1 [2] is our level 2 [1]. The differences are the result of differences in notation between this book and Taguchi's standard notation, as explained in section 3.10.)

A sixteen-run fractional factoral design (Figure 4.33) was replicated four times. The order in which the $16 \times 4 = 64$ trials were performed was randomized as much as possible. The assignment of factors to the columns of Figure 4.33 is given in the last column of Figure 5.19. The observed values of the response (percent post extrusion shrinkage of casings), subsample averages, standard deviations, and S/N ratios are given in Figure 5.20. Since the goal is to minimize shrinkage, the S/N ratio formula for the smaller-is-better case was used. In Figures 5.21, 5.24, and 5.27 the factor effects on mean, log s and S/N ratio are calculated. These estimated effects are presented graphically in Figures 5.22, 5.25, and 5.28. Normal plots of the estimated effects are given in Figures 5.23, 5.26, and 5.29. We should keep in mind that since fifteen effects were estimated using a sixteen-run experimental design, all main effects are confounded with two-factor interaction terms.

Figures 5.21 and 5.22 indicate that factor K has a strong effect on average response. Factor E is next most influential and the rest appear to be random variation. The normal probability plot of effects on average values (Figure 5.23) supports treating factor K as significant. The point for factor A is also away from the fitted line in Figure 5.23. But points which deviate to the right of the lower end of the fitted line or to the left of the upper end of the fitted line indicate estimates which are *closer* to the other estimates than one might expect with normal random variability. These points do not indicate real factor effects.

Based on Figures 5.24 and 5.25, factor M affects variability (log s) the most. (Note that all log s values were less than zero. To increase legibility in Figure 5.24, the minus signs were omitted for the log s values. However, the

Standard order	Observed values				Mean	Standard deviation	S/N ratio	Log s
1	0.58	0.62	0.59	0.54	0.58	0.033	4.68	−1.48
2	0.34	0.32	0.30	0.41	0.34	0.048	9.24	−1.32
3	0.28	0.26	0.26	0.30	0.27	0.019	11.20	−1.72
4	0.13	0.17	0.21	0.17	0.17	0.033	15.27	−1.48
5	0.54	0.53	0.53	0.54	0.53	0.006	5.43	−2.22
6	0.48	0.49	0.44	0.41	0.45	0.037	6.82	−1.43
7	0.07	0.04	0.19	0.18	0.12	0.076	17.27	−1.12
8	0.08	0.10	0.14	0.18	0.12	0.044	17.67	−1.36
9	0.13	0.19	0.19	0.19	0.17	0.030	15.05	−1.52
10	0.24	0.22	0.19	0.25	0.22	0.026	12.91	−1.59
11	0.16	0.17	0.13	0.12	0.14	0.024	16.69	−1.62
12	0.13	0.22	0.20	0.23	0.19	0.045	14.03	−1.35
13	0.16	0.16	0.19	0.19	0.17	0.017	15.11	−1.77
14	0.07	0.09	0.11	0.08	0.09	0.017	21.04	−1.77
15	0.55	0.60	0.57	0.58	0.57	0.021	4.80	−1.68
16	0.49	0.54	0.46	0.45	0.48	0.040	6.26	−1.40

FIGURE 5.20 Experimental data and sample statistics

estimated effects at the bottom of Figure 5.24 and the values in Figure 5.25 have proper signs.) Factors G, D, C, E, H, and I are next most influential, but none seems to have a significant effect. Figure 5.26 suggests that M, G, D, C, and I are from a different distribution of effects than the other effects acting on log s. This dichotomization in the data can be explained in part by noting that factors M, G, C, and I were all set at their low levels when the smallest response value was observed (log s = −2.22, trial 5) and were all at their high levels when the largest response was observed (log s = −1.12, trial 7).

For S/N ratios, Figures 5.27, 5.28, and 5.29 show factors K and E to have strong effects. Quinlan also found factors F, A, C, D, J and M to have *statistically* significant effects on the S/N ratio. He based this conclusion on analysis of variance, which will be discussed in Chapter 8.

The above analyses of Figures 5.21 through 5.29 are summarized in Figure 5.30. In that figure the levels of key factors which will optimize the response are listed. Levels in parentheses showed some significance based on Figures 5.22, 5.25, or 5.28, but the normal plots did not show significant departures from normal variability. The values in the last column followed by asterisks were found to be significant only through analysis of variance. Keep in mind that the objective is to minimize the response average (that is, minimize average amount of shrinkage), reduce log s, and maximize the S/N ratio.

There are only three "conflicts" in the levels recommended in Figure 5.30. For factors C and D, level 1 minimizes log s, but level 2 maximizes the S/N ratio. This is not a serious concern, however, since the inclusion of these two factors was tenuous at best, and we are probably observing random variability in action here. The mixed signals on factor E are more serious, however. Factors K and E are clearly the key variables, based on our data analysis. Should factor E be set at level 2, to minimize log s, or at level 1 to minimize $\bar{y}$ and maximize the S/N ratio? This is a difficult call. If level 2 *significantly*

Factor legend: A: Liner O.D. · B: Liner Die · C: Liner Linespeed · D: Liner Tension · J: Liner Material · K: Wire Braid · L: Liner Temp. · M: Braid Tension · N: Coating Material · O: Melt Temp. · E: Wire Diam. · F: Coating Die · G: Screen Pack · H: Cool Method · I: Line Speed

Random Order Trial No.	Std Order Trial No.	Response Observed Value Y	A:1	A:2	B:1	B:2	C:1	C:2	D:1	D:2	J:1	J:2	K:1	K:2	L:1	L:2	M:1	M:2	N:1	N:2	O:1	O:2	E:1	E:2	F:1	F:2	G:1	G:2	H:1	H:2	I:1	I:2
1		0.58	.58		.58		.58		.58			.58		.58		.58		.58		.58		.58	.58		.58		.58		.58			.58
2		.34	.34		.34		.34			.34		.34		.34	.34			.34	.34		.34		.34			.34		.34		.34	.34	
3		.27	.27		.27			.27	.27			.27	.27			.27	.27			.27	.27			.27	.27			.27		.27	.27	
4		.17	.17		.17			.17		.17		.17	.17		.17		.17		.17			.17		.17		.17	.17		.17			.17
5		.53	.53			.53	.53		.53		.53			.53		.53	.53		.53			.53		.53		.53	.53			.53	.53	
6		.45	.45			.45	.45			.45	.45			.45	.45		.45			.45	.45			.45	.45			.45	.45			.45
7		.12	.12			.12		.12	.12		.12		.12			.12		.12	.12		.12		.12			.12		.12	.12			.12
8		.12	.12			.12		.12		.12	.12		.12		.12			.12		.12		.12	.12		.12		.12			.12	.12	
9		.17		.17	.17		.17		.17		.17		.17		.17			.17		.17		.17		.17		.17		.17	.17		.17	
10		.22		.22	.22		.22			.22	.22		.22			.22		.22	.22		.22			.22	.22		.22			.22		.22
11		.14		.14	.14			.14	.14		.14			.14	.14		.14			.14	.14		.14			.14	.14			.14		.14
12		.19		.19	.19			.19		.19	.19			.19		.19	.19		.19			.19	.19		.19			.19	.19		.19	
13		.17		.17		.17	.17		.17			.17	.17		.17		.17		.17			.17	.17		.17			.17		.17		.17
14		.09		.09		.09	.09			.09		.09	.09			.09	.09			.09	.09		.09			.09	.09		.09		.09	
15		.57		.57		.57		.57	.57			.57		.57	.57			.57	.57		.57			.57	.57		.57		.57		.57	
16		.48		.48		.48		.48		.48		.48		.48		.48		.48		.48		.48		.48		.48		.48		.48		.48
TOTAL		4.61	2.58	2.03	2.08	2.53	2.55	2.06	2.55	2.06	1.94	2.67	1.33	3.28	2.13	2.48	2.01	2.60	2.31	2.30	2.20	2.41	1.75	2.86	2.57	2.04	2.42	2.19	2.34	2.27	2.28	2.33
NUMBER OF VALUES		16	8	8	8	8	8	8	8	8	8	8	8	8	8	8	8	8	8	8	8	8	8	8	8	8	8	8	8	8	8	8
AVERAGE		.29	.32	.25	.26	.32	.32	.26	.32	.26	.24	.33	.17	.41	.27	.31	.25	.33	.29	.29	.28	.30	.22	.36	.32	.26	.30	.27	.29	.28	.29	.29
EFFECT			-.07		.06		-.06		-.06		.09		.24		.04		.08		0		.02		.14		-.06		-.03		-.01		0	

FIGURE 5.21 Response table for sample mean values

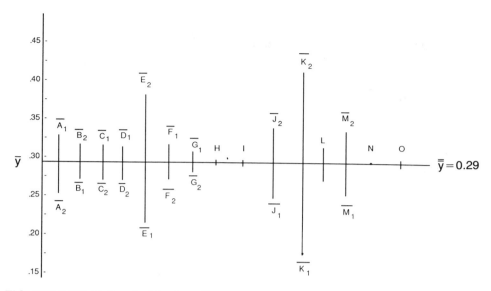

FIGURE 5.22 Estimated factor effects on sample means

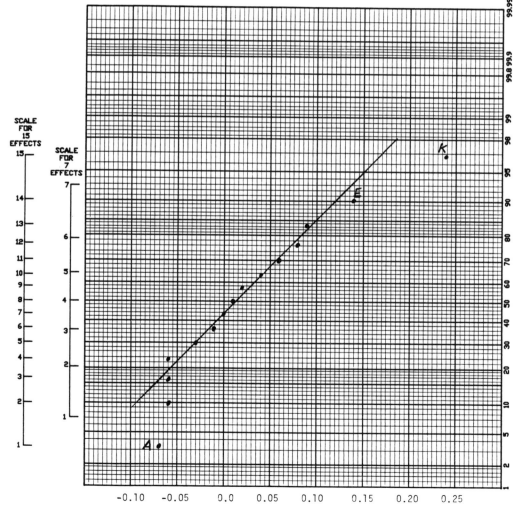

FIGURE 5.23 Normal probability plot for estimated effects on y

Random Order Trial Number	Standard Order Trial Number	Response Observed Values y	A (1)	A (2)	B (1)	B (2)	C (1)	C (2)	D (1)	D (2)	J (1)	J (2)	K (1)	K (2)	L (1)	L (2)	M (1)	M (2)	N (1)	N (2)	O (1)	O (2)	E (1)	E (2)	F (1)	F (2)	G (1)	G (2)	H (1)	H (2)	I (1)	I (2)
	1	1.48	1.48		1.48		1.48		1.48			1.48		1.48		1.48		1.48		1.48		1.48	1.48		1.48		1.48		1.48			1.48
	2	1.32	1.32		1.32		1.32			1.32		1.32		1.32	1.32			1.32	1.32		1.32		1.32			1.32		1.32		1.32	1.32	
	3	1.72	1.72		1.72			1.72	1.72			1.72	1.72			1.72	1.72			1.72	1.72			1.72	1.72			1.72		1.72	1.72	
	4	1.48	1.48		1.48			1.48		1.48		1.48	1.48		1.48		1.48		1.48			1.48		1.48		1.48	1.48		1.48			1.48
	5	2.22	2.22			2.22	2.22		2.22		2.22			2.22		2.22	2.22		2.22			2.22		2.22		2.22	2.22			2.22	2.22	
	6	1.43	1.43			1.43	1.43			1.43	1.43			1.43	1.43		1.43			1.43	1.43			1.43	1.43			1.43	1.43			1.43
	7	1.12	1.12			1.12		1.12	1.12		1.12		1.12			1.12		1.12	1.12		1.12		1.12			1.12		1.12	1.12			1.12
	8	1.36	1.36			1.36		1.36		1.36	1.36		1.36		1.36			1.36		1.36		1.36	1.36		1.36		1.36		1.36		1.36	
	9	1.52		1.52	1.52		1.52		1.52		1.52		1.52		1.52			1.52		1.52		1.52		1.52		1.52		1.52	1.52		1.52	
	10	1.59		1.59	1.59		1.59			1.59	1.59		1.59			1.59		1.59	1.59		1.59			1.59	1.59		1.59			1.59		1.59
	11	1.62		1.62	1.62			1.62	1.62		1.62			1.62	1.62		1.62			1.62	1.62		1.62			1.62	1.62			1.62		1.62
	12	1.35		1.35	1.35			1.35		1.35	1.35			1.35		1.35	1.35		1.35			1.35	1.35		1.35			1.35	1.35		1.35	
	13	1.77		1.77		1.77	1.77		1.77			1.77	1.77		1.77		1.77		1.77			1.77	1.77		1.77			1.77		1.77		1.77
	14	1.77		1.77		1.77	1.77			1.77		1.77	1.77			1.77	1.77			1.77	1.77		1.77			1.77	1.77		1.77		1.77	
	15	1.68		1.68		1.68		1.68	1.68			1.68		1.68	1.68			1.68	1.68		1.68			1.68	1.68		1.68		1.68		1.68	
	16	1.40		1.40		1.40		1.40		1.40		1.40		1.40		1.40		1.40		1.40		1.40		1.40		1.40		1.40		1.40		1.40
TOTAL		24.83	12.13	12.70	12.08	12.75	13.10	11.73	13.13	11.70	12.21	12.62	12.33	12.50	12.18	12.65	13.36	11.47	12.53	12.30	12.25	12.58	11.79	13.04	12.38	12.45	13.20	11.63	11.83	13.00	12.94	11.89
NUMBER OF VALUES		16	8	8	8	8	8	8	8	8	8	8	8	8	8	8	8	8	8	8	8	8	8	8	8	8	8	8	8	8	8	8
AVERAGE		1.55	1.52	1.59	1.51	1.59	1.64	1.47	1.64	1.46	1.53	1.58	1.54	1.56	1.52	1.58	1.67	1.43	1.57	1.54	1.53	1.57	1.47	1.63	1.55	1.56	1.65	1.45	1.48	1.63	1.62	1.49
EFFECT			−0.07		−0.08		0.17		0.18		−0.05		−0.02		−0.06		0.24		0.03		−0.04		−0.16		−0.01		0.20		−0.15		0.13	

FIGURE 5.24 Response table for sample log s

(Note: all values for log s are less than zero; minus signs have been omitted in all but the last row to enhance legibility)

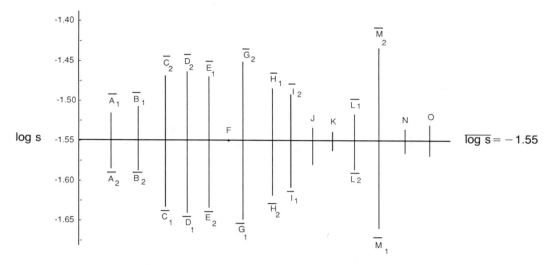

FIGURE 5.25 Estimated factor effects on log(s)

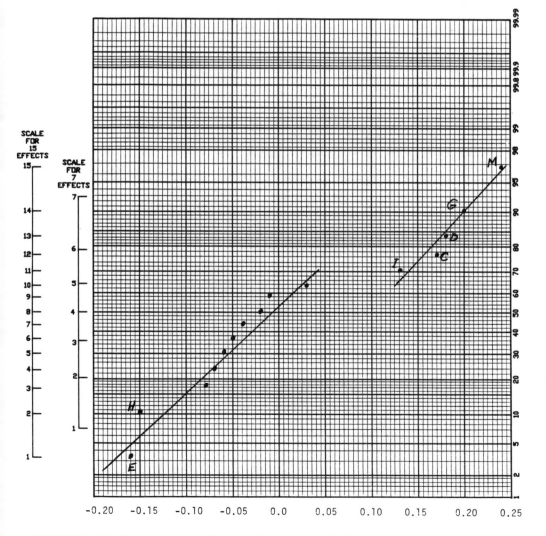

FIGURE 5.26 Normal probability plot for estimated effects on log s

FIGURE 5.27 Response table for S/N ratios

Random Order Trial Number	Standard Order Trial Number	Response Observed Values y	A 1	A 2	B 1	B 2	C 1	C 2	D 1	D 2	J 1	J 2	K 1	K 2	L 1	L 2	M 1	M 2	N 1	N 2	O 1	O 2	E 1	E 2	F 1	F 2	G 1	G 2	H 1	H 2	I 1	I 2
	1	4.68	4.68		4.68		4.68		4.68		4.68			4.68	4.68		4.68		4.68		4.68	4.68		4.68		4.68		4.68		4.68		4.68
	2	9.24	9.24		9.24		9.24		9.24		9.24		9.24	9.24	9.24			9.24	9.24		9.24		9.24		9.24		9.24		9.24		9.24	
	3	11.20	11.20		11.20		11.20		11.20				11.20	11.20			11.20		11.20		11.20	11.20			11.20	11.20			11.20	11.20	
	4	15.27	15.27		15.27			15.27	15.27		15.27		15.27			15.27	15.27			15.27		15.27	15.27		15.27		15.27				15.27
	5	5.43	5.43			5.43	5.43		5.43			5.43			5.43		5.43		5.43		5.43		5.43		5.43		5.43		5.43		5.43	
	6	6.82	6.82			6.82	6.82			6.82	6.82		6.82		6.82		6.82			6.82	6.82			6.82	6.82		6.82		6.82			6.82
	7	17.27	17.27			17.27		17.27	17.27		17.27	17.27	17.67	17.67	17.67		17.27		17.27	17.67	17.27		17.27	17.67	17.67		17.67		17.27		17.27	
	8	17.67	17.67			17.67	17.67		17.27		17.6	17.6	17.67	17.67	15.05		12.9			17.67	17.27		17.67	17.67	12.9		17.67		17.27			7.27
	9	15.05	15.05		15.05		15.05		15.05		15.05	15.05	15.05				15.05		15.05	15.05	12.9		15.05		15.05		15.05		15.05	15.05	15.05	
	10	12.91	12.91	12.91		12.91	12.91			12.91	12.9	12.9	12.91	12.91				12.9	12.91	12.91	12.91			12.91	12.91			12.9	12.91		12.91	
	11	16.69	16.69	16.69	16.69			16.69	16.69		16.69			16.69	16.69	16.69	16.67			16.69	16.69	16.69	16.89			16.69	16.69	16.69		16.69		16.69
	12	14.03	14.03	14.03	14.03			14.03	14.03		14.03	14.03	14.03			14.03	14.03	14.03	14.03				14.03	14.03	14.03			14.03	14.03	14.03	14.03	
	13	15.11	15.11	15.11	15.11		15.11	15.11	15.11			15.11	15.11	15.11	15.11			15.11	15.11	15.11	12.91		15.11	15.11	15.11		15.11		15.11		15.11	
	14	21.04	21.04	21.04	21.04		21.04		21.04		21.04	21.04	21.04	21.04	21.04	21.04	21.04			21.04	21.04	21.04	21.04		21.04		21.04	21.04	21.04	21.04	21.04	
	15	4.80	4.80	4.80		4.80	4.80		4.80		4.80		4.80		4.80	4.80		4.80	4.80	4.80	4.80			4.80		4.80	4.80			4.80	4.80	4.80
	16	6.26	6.26	6.26	6.26		6.26	6.26	6.26	6.26	6.26	6.26	6.26	6.26	6.26	6.26	6.26	6.26	6.26	6.26	6.26	6.26	6.26	6.26	6.26	6.26	6.26	6.26	6.26	6.26	6.26	6.26
TOTAL		192.47	87.95	105.88	99.07	94.40	90.28	103.19	90.23	103.24	105.53	87.60	125.52	67.95	100.46	92.82	105.59	87.88	94.26	99.41	99.97	93.50	115.73	77.74	87.22	106.25	98.49	94.25	98.98	94.98	98.46	95.01
NUMBER OF VALUES		16	8	8	8	8	8	8	8	8	8	8	8	8	8	8	8	8	8	8	8	8	8	8	8	8	8	8	8	8	8	8
AVERAGE		12.09	10.95	13.24	12.38	11.80	11.29	12.90	11.28	12.90	13.23	10.95	15.69	8.49	12.58	11.60	13.20	10.99	11.76	12.43	12.50	11.69	14.47	9.72	10.90	13.28	12.31	11.87	12.37	11.81	12.31	11.88
EFFECT			2.29		-0.58		1.61		1.63		-2.28		-7.20		-0.98		-2.21		0.67		-0.81		-4.75		2.38		-0.44		-0.56		-0.43	

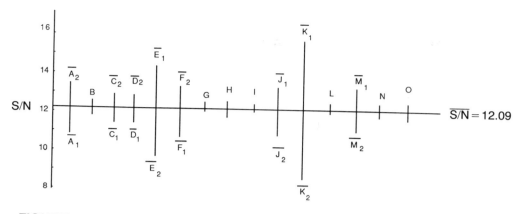

FIGURE 5.28 Estimated factor effects on S/N ratios

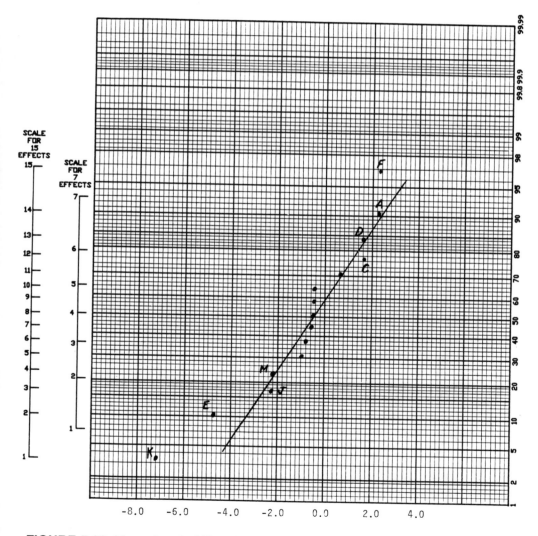

FIGURE 5.29 Normal probability plot for estimated effects on S/N ratio

Factor	$\bar{y}$	log s	S/N ratio
		LEVEL WHICH WILL OPTIMIZE:	
A. Liner O.D.			(2)*
C. Linear line speed		(1)	(2)*
D. Linear tension		(1)	(2)*
E. Wire diameter	(1)	(2)	1
F. Coating die type			(2)*
G. Screen pack		(1)	
H. Cooling method		(2)	
I. Line speed		(1)	
J. Liner material			(1)*
K. Wire braid type	1		1
M. Braiding tension		(1)	(1)*

FIGURE 5.30 Summary of analyses of factor effects

reduced log s, then it would probably be best to go with level 2. After all, if variability could be reduced to almost zero, it would be a simple matter to cut casings a little long to allow for the predictable shrinkage. In this case, however, the reduction in log s does not appear to be significant, but the reductions in $\bar{y}$ and the S/N ratio do appear to be relatively large. So setting factor E at level 1 would appear to be the recommended option. Note that we are not saying that log s is unimportant, but rather that the effect on log s may be simply random variability rather than a real effect.

What should be the conclusion of this study? That factors E and K should both be run at level 1. Confirmatory runs should be performed to verify this decision. Additional experimentation based on the other factors listed in Figure 5.30 should also be considered. Additional data may show some of these other factors to have real effects. Also, the "optimum" levels for factors mentioned in this example refer to the better of only two possible levels for each factor. Other, completely different, factor levels may produce significant improvement.

Since E and K were the two dominant factors in this study, we should consider whether their interaction affected the response. From the listing of confounding relationships for fifteen-factor designs in Section 4.3, we find that the EK interaction is confounded with the B main effect (as well as with some other interaction terms). But factor B had no appreciable effect on response average, standard deviation, or S/N ratio, so we would conclude at this point that their interaction is not important for this study.

S/N ratios?

The signal-to-noise ratio is of some interest, but generally we are better off working instead with both $\bar{y}$ and log s. The ability to distinguish between effects on average and effects on variability helps us make sound, informed decisions. However, this is not to diminish the extremely important contributions of Dr. Taguchi. In particular, Dr. Taguchi made experimenters and statisticians aware of the value in using experimental designs to assess impact of factors on product and process variability. Before Dr. Taguchi came on the scene, most statisticians and experimenters used fractional factorial designs only to assess effect on averages. Now these designs are also used to assess variability, as has been described in this chapter.

6

Taguchi Inner and
Outer Arrays

6.1 Noise Factors

The experimental designs discussed in earlier chapters have focused on selection of levels of factors which would optimize, in some sense, the performance of a product or process. For example, we want to determine what settings for process factors will maximize average strength or minimize variability of coating thickness. Factors which can and should be controlled are called *control factors* by Taguchi and *design parameters* by Kackar (1985).

Some factors which affect product or process performance cannot be controlled or are difficult to control. For example, air temperature or vibration may affect automobile carburetor performance; composition of incoming scrap metal will affect ingots produced in a refining process; humidity in a factory may affect quality characteristics in a coating operation. Factors which affect process or product characteristics but are difficult, expensive, or impossible to control are called *noise factors* by Dr. Taguchi. He identified three types of noise factors (see section 2.3): *external noise*, or variation in environmental conditions, such as dust, temperature, humidity, or supply voltages; *internal noise* or deterioration, such as product wear, material aging, or other changes in components or materials with time or use; and *unit-to-unit noise*, which is differences in products built to the same specifications, and is caused by variability in materials, manufacturing equipment, and assembly processes.

In the past, many engineers tried to deal with problems related to noise factors by controlling the noise factors themselves: they hermetically sealed components sensitive to humidity; isolated components sensitive to vibration; air conditioned operations in which temperature is a key factor, or assured consistency of incoming materials or parts through intense inspection or testing. Dr. Taguchi proposed that such control actions be used only as a last resort, since they are often quite expensive. Instead he recommended that, through experimentation, those levels of design parameters which will *minimize the impact the noise factors have on product and process performance* be identified. The control factors are then set at those levels which will make the product *robust* or insensitive to the noise factors.

6.2 Experimental Designs for Control and Noise Factors

Three approaches have been suggested for designing experiments to analyze both control and noise factors:

> *Approach 1.* Make no attempt to control noise factors during the experiment. Instead, run replicated trials for selected combinations of control factors and measure process variability with sample standard deviations calculated at each experimental "point." The environment in which the experiment is performed should be as close to actual use or manufacture as possible.

> *Approach 2.* Identify noise factors prior to experimentation, and include the noise factors with the control factors in the experimental design. For example, if it were necessary to deal with *six* control factors and *two* noise factors, a sixteen-run experiment in *eight* factors would be used. Interactions between design and noise factors would be evaluated, and levels of control factors which minimized noise factor effects would be identified.

> *Approach 3.* Select an experimental design for the control factors. Taguchi called this design matrix an *inner array*. Select a second experimental design for the noise factors. This is called the *outer array*. For each combination of factors in the inner array, run all the combinations of noise factors in the outer array. (So if the inner array contains eight rows and the outer array contains four rows, then the combined experiment will have $8 \times 4 = 32$ experimental trials).

There has been considerable debate as to the best approach to designing an experiment for considering both control and noise factors. Approach 3 was proposed by Taguchi. It can be viewed as a special case of Approach "2" in which constraints are imposed on the assignment of factors to columns of a design matrix. It has some desirable features regarding confounding relationships. In particular, all the factors in the inner array are orthogonal to all the factors in the outer array. There is no confounding of control factor effects or interactions with noise factor effects or interactions. Further, all two-factor interactions *between control and noise factors* can be estimated free of confounding with control factor effects or interactions, or noise factor effects or interactions. However, Approach 3 can require a large number of experimental runs, and despite the large number of trials, the control factor effects and interactions might still be confounded among themselves. See Lochner and Matar (1988) for further discussion.

As an example, consider an experimental situation involving five control factors and three noise factors. Suppose an eight-run design is selected for the inner array and a four-run experiment for the outer array. You will need $8 \times 4 = 32$ experimental trials. The following confounding relationships would occur in this thirty-two-run experiment:

- All main effects for control factors are confounded with two-factor control interactions (see section 4.1).
- The design matrix for three factors in four runs is obtained by setting C equal to AB in the factorial design for two factors in four runs, producing the design matrix in Figure 6.1 (see also Figures 3.46 and 3.47). Each of the three noise factors would be confounded with a two-factor noise interaction (which, by the way, is generally not viewed as a serious problem when considering noise factors).

Consider now using Approach 2 for handling five control factors and three noise factors. There are a total of $5 + 3 = 8$ factors. From section 4.3, a sixteen-run experiment in eight factors allows estimation of all main effects independent of two-factor interactions.

Bradley Jones, Chief Scientist at Catalyst, Inc., makes the following observations regarding use of outer arrays (1988):

Typically, noise factors are costly to control. However, American manufacturers often must control these factors in order to reduce the variability of the response. This increases operating costs of processes. By contrast, firms that discover operating regions within which the response is relatively unaffected by changes in the noise factors will enjoy a competitive advantage. In a Taguchi-style experiment the noise factors are often rigidly controlled in an outer array. This is expensive, but worth it if the experiment is successful.

The outer array may be any one of the commonly used orthogonal arrays. The key point here is that the outer arrays are *orthogonal*. In an orthogonal design, the factor effects are independent of one another which is often desirable. However, orthogonality in the outer array also implies that the noise factors vary *independently* in reality. This is an unwarranted and dangerous assumption. It is unwarranted

Standard order	FACTORS		
	$A = BC$	$B = AC$	$C = AB$
1	1	1	2
2	1	2	1
3	2	1	1
4	2	2	2

FIGURE 6.1 Design matrix for three factors in four trials

because process variables often move in tandem, not independently. It is dangerous because this correlation in the noise factors can mean that the variance of the responses over the outer array may either overestimate or underestimate the true variance of the response. . . .

Thus, in practice, if the noise factors are correlated, then the measures of the variability over the outer arrays may overestimate the noise for some inner array points and underestimate it for others. It would be impossible to reliably find the setting of the control factors that minimizes the noise by using outer arrays. An alternative is to repeat the inner-array point as many times as an outer array would require, but to allow the noise factors to vary randomly during the course of the experiment. [Note: This is Approach 1 above.] In following this strategy, total randomization is prudent. In this way the natural variability of the noise factors should produce a better behaved estimate of variability in the response.

Another alternative would be to use an historical study to determine the correlations of the noise factors in process, then continue to use an orthogonal array for the outer arrays, but use a weighting scheme on the outer array points when forming the performance statistics (signal-to-noise ratios) of interest.

(This last possibility will not be pursued here, in order to avoid adding complexity to the discussion on experimental designs.)

The authors of this book recommend a combination of Approaches 1 and 2. Noise factors which are considered critical should be treated just like control factors whenever possible. Variability due to the noise factors which are not being controlled can be evaluated by replicating the experiment and calculating sample standard deviations at each experimental point.

6.3 Illustrative Example (Example 6.1)

The example in this section is based on a study by Guy Desrochers and Dale Ewing of Eaton Yale Limited (1984). It is also discussed in Pignatiello and Ramberg (1985). The purpose of the experiment was to reduce the variability in unloaded camber height of leaf springs manufactured by the Suspension Division of Eaton Yale Limited. It was also desired to have the average height measurements as close as possible to a specified nominal value of 8.0.

There are four control factors and one noise factor in the experiment. The inner array is an eight-run design and the outer array a two-run design (that is, levels 1 and 2 for the noise factor). Each of the $8 \times 2 = 16$ combinations of factors used in the experiment was replicated three times, for a total of $16 \times 3 = 48$ trials. Desrochers and Ewing based their analysis on average response values and signal-to-noise ratios. We will use $\bar{y}$ and $log\ s$ as the *performance statistics*, and will also discuss Desrochers and Ewing's results using the S/N ratio.

The four key process variables (control factors) identified for the experiment were:

Furnace temperature, the temperature setting for the heat treatment furnace;

Heating time, the length of time the part is heated;

Transfer time, the time it takes a part to travel from the furnace to the camber former; and

Hold-down time, the length of time the camber former is closed on a part.

In addition, the temperature of the oil used to quench parts was included as an (uncontrollable) noise factor. The experiment actually performed involved an eight-run inner array and a two-run outer array, but can also be described as a sixteen-run experiment in five factors, replicated three times. The columns in the design matrix (Figure 4.33) to which the factors were assigned, and the experimental levels for the factors, are given in Figure 6.2. In section 4.3 it was recommended that, when analyzing five factors in sixteen runs, the fifth factor be assigned to column 15. But in Desrochers and Ewing's study, the fifth factor was assigned to column 14 instead. The resulting assignment of two-factor interaction terms can be easily determined by referring in section 4.3 to the alias structure for eight factors and crossing out all interactions involving factors E, F, or G. The interaction columns are identified in Figure 6.3. Note in Figure 6.3 that some of the two-factor interactions are confounded. (If the design recommended in section 4.3 for use with five factors had been used, this confounding would have been avoided.)

The factor combinations used, the observed values for the response (unloaded camber height) and calculated values for $\bar{y}$, s, and $log\ s$ are listed in Figure 6.4. Figures 6.5 and 6.7 are the completed response tables for the $\bar{y}$ and $log\ s$ values. (The minus signs in front of the $log\ s$ values are omitted in Figure 6.7, but the estimated effects on the bottom line have the correct sign.) Figures 6.6 and 6.8 are normal probability plots of the effects estimated in Figures 6.5 and 6.7, respectively.

Figures 6.5 and 6.6 indicate that main effects A, B, and C, and the interaction term AC may have significant effects on the average response. Figure 6.8 shows only factor B having any real effect on $log\ s$. Because of possible AC interaction effects, the average response should be estimated at each combination of levels of A and C. This is done in Figure 6.9, and the interaction is presented graphically in Figure 6.10. Figure 6.10 shows a clear interaction effect. In order to get the average response as close as possible to the target value (which is 8.0), factors A and C should both be set at level 1, where the estimated mean effect is 7.94. However, if A is a noise factor which cannot be controlled, it might be best to set C at level 2 to minimize the variability in the average response.

Factor B affects both response average and variability. In order to get the average response closer to the target value, B should be set at level 2. But to reduce variability of the response variable, B should be set at level 1. This is a problem. Given the important impact factor B has on variability, it might be best to set it at level 1 and seek alternate ways of adjusting the average response closer to target, such as further adjustment of factors A and C. If we set factors A, B, and C at level 1, the estimated average response is $7.94 + (7.53 - 7.64) = 7.83$, which is comfortably close to 8.0.

	FACTOR		FACTOR LEVELS	
Column	Code	Name	Low	High
1	A	Quench oil temperature	130°-150°F	150°-170°F
2	B	Furnace temperature	1840°F	1880°F
3	C	Heating time	25 sec	23 sec
4	D	Transfer time	12 sec	10 sec
14	H	Hold down time	2 sec	3 sec

FIGURE 6.2 Experimental levels for example factors

Column	Interactions
5	AB
6	AC
7	AD
8	BC = DH
9	BD = CH
10	BH = CD
15	AH

FIGURE 6.3 Alias structure for two-factor interactions

Standard order	A	B	C	D	H	Response values			$\bar{y}$	s	log s
1	1	1	1	1	1	7.78	7.78	7.81	7.79	0.017	−1.76
2	1	1	1	2	2	7.94	8.00	7.88	7.94	0.060	−1.22
3	1	1	2	1	2	7.50	7.56	7.50	7.52	0.035	−1.46
4	1	1	2	2	1	7.56	7.62	7.44	7.54	0.092	−1.04
5	1	2	1	1	2	8.15	8.18	7.88	8.07	0.165	−0.78
6	1	2	1	2	1	7.69	8.09	8.06	7.95	0.223	−0.65
7	1	2	2	1	1	7.59	7.56	7.75	7.63	0.102	−0.99
8	1	2	2	2	2	7.56	7.81	7.69	7.69	0.125	−0.90
9	2	1	1	1	1	7.50	7.25	7.12	7.29	0.193	−0.71
10	2	1	1	2	2	7.32	7.44	7.44	7.40	0.069	−1.16
11	2	1	2	1	2	7.50	7.56	7.50	7.52	0.035	−1.46
12	2	1	2	2	1	7.18	7.18	7.25	7.20	0.040	−1.39
13	2	2	1	1	2	7.88	7.88	7.44	7.73	0.254	−0.60
14	2	2	1	2	1	7.56	7.69	7.62	7.62	0.065	−1.19
15	2	2	2	1	1	7.63	7.75	7.56	7.65	0.096	−1.02
16	2	2	2	2	2	7.81	7.50	7.59	7.63	0.160	−0.80

FIGURE 6.4 Data and performance statistics for example

Random Order Trial Number	Standard Order Trial Number	Response Observed Value y	A: Quench Temp		B: Furnace Temp		C: Heating Time		D: Transfer Time		AB		AC		AD		BC=DH		BD=CH		BH=CD								H: Hold Down Time		AH	
			1	2	1	2	1	2	1	2	1	2	1	2	1	2	1	2	1	2	1	2	1	2	1	2	1	2	1	2	1	2
1		7.79	7.79		7.79		7.79		7.79					7.79	7.79	7.79		7.79	7.79		7.79		7.79		7.79		7.79		7.79		7.79		7.79
2		7.94	7.94		7.94		7.94		7.94				7.94 7.94		7.94	7.94		7.94		7.94		7.94		7.94	7.94		7.94		7.94	7.94		7.94	
3		7.52	7.52		7.52		7.52 7.52		7.52		7.52 7.20		7.52			7.52	7.52			7.52		7.52 7.52	7.52		7.52	7.52	7.52		7.52 7.52		7.52 7.52	
4		7.54	7.54		7.54		7.54			7.54	7.54 7.20		7.54 7.54		7.54	7.54		7.54		7.54		7.54		7.54	7.54	7.54		7.54		7.54		7.54	
5		8.07	8.07			8.07	8.07		8.07		8.07			8.07	8.07		8.07	8.07		8.07		8.07		8.07		8.07		8.07		8.07		8.07	
6		7.95	7.95		7.63		7.95		7.95		7.95		7.95			7.95	7.95		7.95		7.95	7.45		7.45		7.45		7.95		7.95		7.95	
7		7.63	7.63		7.63		7.63		7.63		7.63		7.62	7.63		7.63	7.63	7.63	7.69		7.63	7.63		7.63	7.63		7.63	7.63		7.63		7.63	
8		7.69	7.69		7.69		7.69	7.69	7.69		7.89		7.69		7.69	7.68		7.69		7.69	7.69		7.69		7.69		7.69		7.69		7.69		7.63
9		7.29	7.29	7.29	7.29		7.29		7.29		7.29		7.29		7.24	7.29		7.29	7.29		7.29		7.29		7.29		7.29		7.29		7.29		
10		7.40	7.40	7.40	7.40		7.40		7.40		7.40		7.40		7.40	7.40		7.40		7.40		7.40		7.40		7.40		7.40		7.40		7.40	
11		7.52	7.52	7.52	7.52		7.52 7.52		7.52 7.52		7.52			7.52	7.52	7.52		7.52 7.52		7.52 7.52		7.52	7.52		7.52		7.52		7.52			7.52
12		7.20	7.20	7.20	7.20		7.20		7.20		7.20		7.20		7.20	7.20		7.20		7.20		7.20	7.20		7.20		7.20		7.20	7.20		7.20	
13		7.73	7.73	7.73	7.73		7.73		7.73			7.73	7.73		7.73	7.73		7.73		7.73		7.73	7.73		7.73		7.73		7.73		7.73		7.73
14		7.62	7.62	7.62	7.62		7.62		7.62			7.62	7.62		7.62 7.62			7.62	7.62		7.62		7.62	7.62		7.62	7.62		7.62		7.62		
15		7.65	7.65	7.65	7.65	7.65	7.65	7.65	7.65	7.65	7.65	7.65			7.65	7.65			7.65	7.65		7.65	7.95	7.95		7.65		7.65	7.65		7.65		
16		7.63	7.63	7.63	7.63	7.63	7.63	7.63	7.63	7.63	7.63	7.63	7.63	7.63	7.63	7.63	7.63	7.63	7.63	7.63	7.63	7.63	7.63	7.63	7.63	7.63	7.63	7.63	7.63	7.63	7.63	7.63	7.63
TOTAL		122.17	62.13	60.04	60.20	61.97	61.79	60.38	61.20	60.97	60.75	61.42	60.42	61.75	61.31	60.86	61.15	61.02	61.16	61.01	61.23	60.94	61.12	61.05	60.93	61.24	61.28	60.89	60.67	61.50	60.98	61.19
NUMBER OF VALUES		16	8	8	8	8	8	8	8	8	8	8	8	8	8	8	8	8	8	8	8	8	8	8	8	8	8	8	8	8	8	8
AVERAGE		7.64	7.77	7.51	7.53	7.75	7.72	7.55	7.65	7.62	7.59	7.68	7.55	7.72	7.66	7.61	7.64	7.63	7.65	7.63	7.65	7.62	7.64	7.63	7.62	7.66	7.66	7.61	7.58	7.69	7.62	7.65
EFFECT			-0.26		0.22		-0.17		-0.03		0.09		0.17		-0.05		-0.01		-0.02		-0.03		-0.01		0.04		-0.05		0.11		0.03	

FIGURE 6.5 Response table for average response

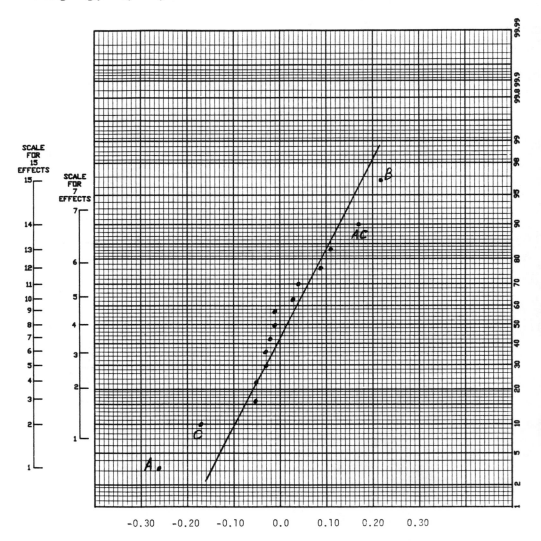

FIGURE 6.6 Normal probability plot of estimated effects on average response

FIGURE 6.7 Response table for log s

(Note: All log s values are below zero; minus signs have been omitted for legibility in all but the final row where signs are included as appropriate)

Random Order Trial Number	Standard Order Trial Number	Response Observed Value y	A: Quench Temp.		B: Furnace Temp.		C: Heating Time		D: Transfer Time		AB		AC		AD		BC=DH		BD=CH		BH=CD								H: Hold Down Time		AH	
			1	2	1	2	1	2	1	2	1	2	1	2	1	2	1	2	1	2	1	2	1	2	1	2	1	2	1	2	1	2
	1	1.76	1.76		1.76		1.76		1.76			1.76		1.76		1.76		1.76		1.76		1.76	1.76		1.76		1.76		1.76			1.76
	2	1.22	1.22		1.22		1.22			1.22		1.22		1.22	1.22			1.22	1.22		1.22		1.22			1.22		1.22		1.22	1.22	
	3	1.46	1.46		1.46			1.46	1.46			1.46	1.46			1.46	1.46			1.46	1.46			1.46	1.46			1.46		1.46	1.46	
	4	1.04	1.04		1.04			1.04		1.04		1.04	1.04		1.04		1.04		1.04			1.04		1.04		1.04	1.04		1.04			1.04
	5	0.78	0.78			0.78	0.78		0.78		0.78			0.78		0.78	0.78		0.78			0.78		0.78	0.78			0.78	0.78		
	6	0.65	0.65			0.65	0.65			0.65	0.65			0.65	0.65		0.65			0.65	0.65			0.65	0.65			0.65	0.65			0.65
	7	0.99	0.99			0.99		0.99	0.99		0.99		0.99			0.99	0.99		0.99		0.99		0.99			0.99		0.99	0.99			0.99
	8	0.90	0.90			0.90		0.90		0.90	0.90		0.90		0.90			0.90		0.90		0.90	0.90		0.90		0.90			0.90	0.90	
	9	0.71		0.71	0.71		0.71		0.71		0.71		0.71		0.71			0.71		0.71		0.71		0.71		0.71		0.71	0.71		0.71	
	10	1.16		1.16	1.16		1.16			1.16	1.16		1.16			1.16		1.16	1.16		1.16			1.16	1.16		1.16			1.16		1.16
	11	1.46		1.46	1.46			1.46	1.46		1.46			1.46	1.46		1.46			1.46	1.46		1.46		1.46		1.46			1.46		1.46
	12	1.39		1.39	1.39			1.39		1.39	1.39			1.39		1.39	1.39		1.39			1.39	1.39		1.39			1.39	1.39		1.39	
	13	0.60		0.60		0.60	0.60		0.60			0.60	0.60		0.60		0.60		0.60			0.60	0.60		0.60			0.60		0.60		0.60
	14	1.19		1.19		1.19	1.19			1.19		1.19	1.19			1.19	1.19			1.19	1.19		1.19		1.19		1.19		1.19		1.19	
	15	1.02		1.02		1.02		1.02	1.02			1.02		1.02	1.02			1.02	1.02		1.02			1.02	1.02		1.02		1.02		1.02	
	16	0.80		0.80		0.80		0.80		0.80		0.80		0.80		0.80		0.80		0.80		0.80		0.80		0.80		0.80		0.80		0.80
TOTAL		17.13	8.80	8.33	10.20	6.93	8.07	9.06	8.78	8.35	8.04	9.09	8.05	9.08	7.60	9.53	8.57	8.56	8.20	8.93	9.15	7.98	9.51	7.62	8.94	8.19	9.31	7.82	8.75	8.38	8.67	8.46
NUMBER OF VALUES		16	8	8	8	8	8	8	8	8	8	8	8	8	8	8	8	8	8	8	8	8	8	8	8	8	8	8	8	8	8	8
AVERAGE		1.07	1.10	1.04	1.28	0.87	1.01	1.13	1.10	1.04	1.01	1.14	1.01	1.14	0.95	1.19	1.07	1.07	1.03	1.12	1.14	1.00	1.19	0.95	1.12	1.02	1.16	0.98	1.09	1.05	1.08	1.06
EFFECT			0.06		0.41		-0.12		0.06		-0.13		-0.13		-0.24		0		-0.09		0.14		0.24		0.10		0.18		0.04		0.02	

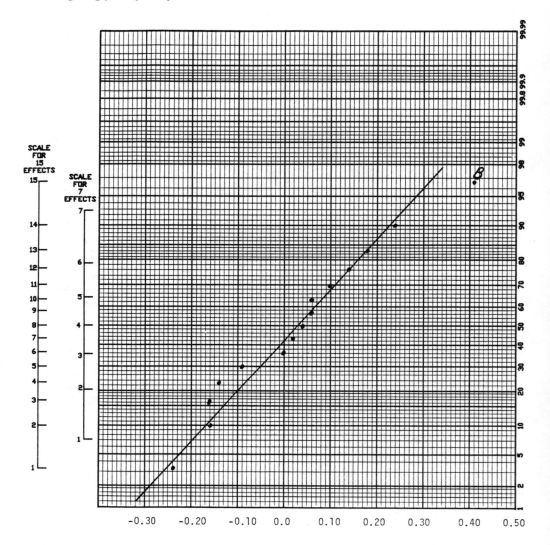

FIGURE 6.8 Normal probability plot of estimated effects on log s

FACTOR LEVEL			
A	*C*	*Sample averages*	*Combined average*
1	1	7.79, 7.94, 8.07, 7.95	7.94
1	2	7.52, 7.54, 7.63, 7.69	7.60
2	1	7.29, 7.40, 7.73, 7.62	7.51
2	2	7.52, 7.20, 7.65, 7.63	7.50

FIGURE 6.9 Average response levels by factor combinations

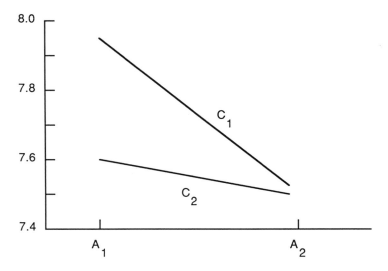

FIGURE 6.10 AC interaction

In their original paper, Desrochers and Ewing analyzed the data using average values and S/N ratios. They recommended setting A and C at level 1 and B at level 2. In a confirmatory experiment they found that "the high significance of Variable C noticed in the original [analysis] was not present in the confirmation experiments." But factor A was not controlled during the confirmation study, and so interaction effects could have clouded the true impact of factor C. For further discussion on this case study, see Pignatiello and Ramberg (1985).

7

Experimental Designs for Factors at Three and Four Levels

7.1 Why Use More Than Two Levels?

In the experimental designs presented in earlier chapters, two levels were used for each factor included in the experiment. Most of the examples included factors which could be set at many different levels (for example, time, temperature, or concentration). For such factors, two "appropriate" levels were selected for use in the experiment. This is often an appropriate way to proceed, particularly during the early stages of an investigation when many factors need to be investigated simultaneously. But situations do arise when three or more factor levels are required. Three common situations are:

1. A factor naturally has three or more definite levels—three shifts, four engine models, five production lines;
2. The experiment must, because of limiting circumstances, be done in three or more blocks; or
3. A process is being fine tuned, and response values which may not be linear over some interval of factor levels need to be analyzed in order to optimize factor settings.

The first two situations call for special experimental designs. A few of the most useful designs will be presented in this chapter. The third situation is better handled using response surface methodology than with the experimental designs presented here. Response surface methods allow the experimenter to evaluate factor interactions and the "shape" of the response surface simultaneously. Because this is an introductory text on experimental design, response surfaces will not be discussed here. The topic is not all that difficult, but to introduce it adequately would require extended discussion. Box and Draper (1987) is an excellent reference on response surface methodology.

The focus of this chapter will be on experimental designs for factors at three and four levels, particularly designs which have been recommended by Dr. Taguchi. Keep in mind that designs with three or more factor levels should generally be avoided during initial experimentation because they require significantly more experimental trials, often have confounding relationships which are difficult to identify, and do not include easily understood interaction effects. In this *introductory* book on experimental designs, the number of different designs has been kept short. For information on broader possibilities for statistically designed experiments, see Box, Hunter, and Hunter (1978).

7.2 Factors at Four Levels

Dealing with a factor at four levels is handled in basically the same way as breaking an experiment into four blocks. The one difference is that interaction terms are generally considered unimportant when blocking. We begin with a design matrix for factors at two levels. Two columns of the design matrix are "assigned" to the four-level factor. The four combinations of 1s and 2s in these columns then determine the levels for the four-level factor. For example, if experimental levels for four-level factor W were to be determined by the values in the columns for factors A and B, then the levels for factor W could be determined by the scheme shown in Figure 7.1.

In addition to the columns for factors A and B, the column for the AB interaction effect is also part of the design. If the levels for A and B are known, it is easy to determine the level for AB. Similarly, if the levels for A and AB are known, then the level for B is easily found. So AB is really part of the W *main effect*, rather than a two-factor interaction. This means that the ABC interaction is now a two-factor interaction, rather than a three-factor interaction. But the meaning of "interaction" is less clear, intuitively, once we go beyond two levels for factors. The moral here is clear—avoid factors at more than two levels whenever possible during early experimentation. If it is necessary to use four levels, watch out for interactions by examining average response values at different combinations of factor levels. If interactions involving factors being analyzed at more than two levels is a real concern, find a statistician or more advanced book on experimental design to help you.

Analysis for four-level factors is reasonably straightforward. Ignore the

A	B	W
1	1	1
1	2	2
2	1	3
2	2	4

FIGURE 7.1 Four-level assignment

three columns related to the four-level factor when doing the arithmetic for the response table. Instead, calculate the average response at each level of the factor and plot it. Analysis of variance (see Chapter 8) may be more appropriate than normal probability plots for analyzing the importance of four-level factors.

Example 7.1

A metal fabrication plant wanted to extend the life of drill bits used by assembly line workers. Six factors were identified as being most important to tool life:

- type of drill bit
- size of drill bit
- drill speed
- material being drilled
- type (manufacturer) of drill
- operator

The first four factors could be reasonably tested at two levels each, but four brands of drills were in use at the plant, and because of differences in handling of tools by operators, it was felt that four line workers should be included in the experiment. The levels of the factors are listed in Figure 7.2. A design matrix with at least ten columns is needed for the experiment: one column for each of the first four factors, and three columns for each of the last two factors. A sixteen-run experiment would be appropriate here.

In section 4.3 it was recommended that, for ten factors in sixteen runs, the factors be assigned to columns 1, 2, 3, 4, 5, 11, 12, 13, 14, and 15 of Figure 4.33. Then the confounding among factors is as indicated in Figure 7.3. Suppose we were to assign factors a, b, c, and d in Figure 7.2 to the first four columns of Figure 4.33, factor e to columns 5 and 11, and factor f to columns 12 and 13. This sounds reasonable, but leads to serious confounding. In particular, the "interaction" of columns 5 and 11 ("EJ"), which is now a main effect, is confounded with c in column 3. A safer approach might be to assign columns to factors e and f first, and then assign columns for a, b, c, and d. If we assign columns 1 and 2 to factor e, then column 5 ("AB") is also assigned

| | LEVEL | | | |
Factor	*1*	*2*	*3*	*4*
a. Type of bit	new	resharpen		
b. Size of bit	0.125 in	0.25 in		
c. Drill speed	400 rpm	800 rpm		
d. Material	soft	hard		
e. Manufacturer	Dreamer	Sandvik	Chicago	Milwaukee
f. Operator	Norm	Kristi	Joe	Ted

FIGURE 7.2 Levels for experimental factors

Column	Alias relationships
1	A = BJ = HI
2	B = AJ = GI
3	C = EJ = FI
4	D = EI = FJ
5	J = AB = CE = DF = GH
6	AC = BE = DG = FH
7	AD = BF = CG = EH
8	AE = BC = DH = FG
9	AF = BD = CH = EG
10	AG = BH = CD = EF = IJ
11	E = CJ = DI
12	F = CI = DJ
13	G = BI = HJ
14	H = AI = GJ
15	I = AH = BG = CF = DE

FIGURE 7.3 Confounding with ten factors in sixteen runs

to e. If we assign columns 3 and 4 to factor f, then column 10 ("CD") is also assigned to f. Factors a, b, c, and d could then be assigned to columns 11, 12, 13, and 14. With this arrangement, no main effects will be confounded with each other. The design matrix for the experiment is given in Figure 7.4. The values in columns 1, 2, 3, 4, 5, and 10 of Figure 4.33 are also included in that figure for reference.

The completed response table for the experiment, using hypothetical data, is given in Figure 7.5. Based on that table, it appears that drill speed and

Standard order	e	1	2	5	f	3	4	10	a 11	b 12	c 13	d 14
1	1	1	1	2	1	1	1	2	1	1	1	1
2	1	1	1	2	2	1	2	1	1	2	2	2
3	1	1	1	2	3	2	1	1	2	1	2	2
4	1	1	1	2	4	2	2	2	2	2	1	1
5	2	1	2	1	1	1	1	2	2	2	1	2
6	2	1	2	1	2	1	2	1	2	1	2	1
7	2	1	2	1	3	2	1	1	1	2	2	1
8	2	1	2	1	4	2	2	2	1	1	1	2
9	3	2	1	1	1	1	1	2	2	2	2	1
10	3	2	1	1	2	1	2	1	2	1	1	2
11	3	2	1	1	3	2	1	1	1	2	1	2
12	3	2	1	1	4	2	2	2	1	1	2	1
13	4	2	2	2	1	1	1	2	1	1	2	2
14	4	2	2	2	2	1	2	1	1	2	1	1
15	4	2	2	2	3	2	1	1	2	1	1	1
16	4	2	2	2	4	2	2	2	2	2	2	2

FIGURE 7.4 Design matrix for example

Standard Order Trial Number	Response Observed Value Y	e		e		f		f		e		f											f		Type of bit new / re-sharp		Size of bit 1/8 / 1/4		Drill speed 400 / 800		Material soft / hard	
		1	2	1	2	1	2	1	2	1	2	1	2	1	2	1	2	1	2	1	2	1	2	1	2	1	2	1	2	1	2
1	16.4	16.4		16.4		16.4		16.4		16.4		16.4		16.4		16.4		16.4		16.4		16.4		16.4		16.4		16.4		16.4		16.4
2	11.5	11.5		11.5			11.5	11.5		11.5		11.5		11.5		11.5		11.5			11.5	11.5		11.5			11.5	11.5		11.5		11.5
3	7.7	7.7		7.7		7.7	7.7		7.7	7.7		7.7	7.7	7.7	7.7	7.7		7.7	7.7	7.7		7.7		7.7	7.7	7.7	7.7		7.7		7.7	
4	18.6	18.6			18.6	18.6		18.6			18.6	18.6		18.6		18.6		18.6	18.6		18.6	18.6		18.6	18.6		18.6		18.6		18.6
5	14.1	14.1		14.1		14.1			14.1	14.1		14.1		14.1		14.1		14.1	14.1		14.1	14.1		14.1	14.1	14.1		14.1		14.1	
6	13.1	13.1			13.1	13.1			13.1	13.1		13.1	13.1	13.1			13.1	13.1	13.1	13.1			13.1	13.1	13.1	13.1		13.1		13.1	
7	12.3	12.3		12.3			12.3	12.3			12.3	12.3		12.3		12.3		12.3		12.3	12.3			12.3	12.3	12.3	12.3		12.3		12.3
8	15.5	15.5		15.5		15.5			15.5	15.5		15.5	15.5	15.5	15.5	15.5		15.5		15.5	15.5	15.5		15.5	15.5	15.5	15.5		15.5	15.5	
9	16.1	16.1			16.1	16.1		16.1		16.1		16.1	16.1	16.1	16.1	16.1		16.1	16.1	16.1	16.1			16.1	16.1		16.1		16.1		16.1
10	14.8	14.8		14.8			14.8		14.8	14.8		14.8	14.8	14.8		14.8	14.8		14.8		14.8	14.8		14.8		14.8	14.8		14.8		14.8
11	12.4	12.4		12.4		12.4			12.4		12.4	12.4	12.4	12.4	12.4	12.4		12.4		12.4		12.4		12.4	12.4	12.4		12.4		12.4
12	12.9	12.9		12.9		12.9		12.9			12.9	12.9	12.9	12.9		12.9		12.9	12.9	12.9		12.9		12.9	12.9	12.9		12.9		12.9	
13	11.2	11.2		11.2		11.2	11.2		11.2		11.2	11.2		11.2	11.2	11.2		11.2	11.2	11.2		11.2		11.2	11.2	11.2	11.2		11.2		11.2
14	17.1	17.1		17.1		17.1	17.1		17.1		17.1	17.1	17.1	17.1		17.1		17.1	17.1	17.1			17.1	17.1	17.1	17.1	17.1		17.1		17.1
15	17.5	17.5		17.5		17.5	12.5	17.5		17.5		17.5	17.5	17.5		17.5		17.5	17.5	17.5		17.5		17.5	17.5	17.5	17.5		17.5		17.5
16	10.3	10.3	10.3	10.3	10.3	10.3	10.3	10.3	10.3	10.3	10.3	10.3	10.3	10.3	10.3	10.3	10.3	10.3	10.3	10.3	10.3	10.3	10.3	10.3	10.3	10.3	10.3	10.3	10.3	10.3	10.3
TOTAL	221.5	109.2	112.3	110.4	111.1	114.3	107.2	107.7	113.8	112.9	108.6	107.1	114.4	115.9	105.6	108.2	113.3	112.9	108.6	106.4	115.1	109.3	112.2	108.1	112.4	109.3	112.2	126.4	95.1	129.0	97.5	112.4	109.2
NUMBER OF VALUES	16	8	8	8	8	8	8	8	8	8	8	8	8	8	8	8	8	8	8	8	8	8	8	8	8	8	8	8	8	8	8	8	8
AVERAGE	13.84	13.65	14.04	13.80	13.89	14.29	13.40	13.46	14.23	13.90	13.79	13.39	14.30	14.49	13.20	13.53	14.16	14.11	13.58	13.30	14.39	13.66	14.03	13.64	14.05	13.66	14.05	15.80	11.89	15.50	12.19	14.05	13.84
EFFECT		0.39		0.09		−0.89		0.77		−0.11		−0.63		−1.29		0.91		−0.53		1.09		0.37		0.41		−3.41		−3.31		−0.41		

FIGURE 7.5 Response table for average drill bit life

Manufacturer	Average response	Operator	Average response
1. Dreamer	13.55	1. Norm	14.45
2. Sandvik	13.75	2. Kristi	14.13
3. Chicago	14.05	3. Joe	12.48
4. Milwaukee	14.03	4. Ted	14.33

FIGURE 7.6 Average drill bit life, by manufacturer and operator

material have a significant effect on average drill bit life, with bits lasting longer when the drill speed is 400 rpm and when soft material is being drilled. In order to estimate possible effects of manufacturer and operator, average responses at each level of each of these two factors were calculated. For example, since by Figure 7.4, Manufacturer 1 was used in the first four runs, the average response for Manufacturer 1 is the average of the first four response values in Figure 7.5. These average responses are given in Figure 7.6. Figure 7.6 suggests that there is no real difference between manufacturers, but Joe may be harder on bits than the other three operators.

7.3 Factors at Three Levels

Three different designs will be considered in this section for factors at three levels:

- L_9, a nine-run experimental design for up to four factors at three levels each;
- L_{18}, an eighteen-run experimental design for up to seven factors at three levels each plus one factor at two levels;
- L_{27}, a twenty-seven-run experimental design for up to thirteen factors at three levels each.

The design matrices for these three designs are given in Figures 7.7, 7.8 and 7.9. There are many other experimental designs possible for factors at three levels, but these should be adequate for many applications.

All three designs are *orthogonal*. That is, for any pair of columns, if we:

1. replace the 1s, 2s and 3s by -1s, 0s, and 1s,
2. multiply the corresponding row terms from the two columns together and
3. add up the resulting products,

the sum of these products will always be zero. (If each level of a factor appears the same number of times in a design matrix, and if each level of a factor appears with each level of any other factor the same number of times, then the design is orthogonal.) The importance of orthogonality is that it allows us to estimate each main effect without confounding with other main effects. Alias relationships with interaction effects may be present, however,

Standard	COLUMN			
order	1	2	3	4
1	1	1	1	1
2	1	2	2	2
3	1	3	3	3
4	2	1	2	3
5	2	2	3	1
6	2	3	1	2
7	3	1	3	2
8	3	2	1	3
9	3	3	2	1

FIGURE 7.7 Design matrix for L_9 experiment

and may mislead us. If it is suspected that interactions may be present, then the argument for starting the study with two-level designs and later going to response surface methodology is very persuasive.

When dealing with factors at three levels, the simplest approach to analysis is to calculate the average value at each of the three levels, and plot them on the same graph. Analysis of variance (Chapter 8) may replace normal probability plots as the method of identifying factors which significantly affect the response variable. Response tables which can be used to simplify calculation of average values at different factor levels for L_9, L_{18}, and L_{27} are given in Figures 7.10, 7.11, and 7.12 respectively. The format for these re-

Standard	COLUMN							
order	1	2	3	4	5	6	7	8
1	1	1	1	1	1	1	1	1
2	1	1	2	2	2	2	2	2
3	1	1	3	3	3	3	3	3
4	1	2	1	1	2	2	3	3
5	1	2	2	2	3	3	1	1
6	1	2	3	3	1	1	2	2
7	1	3	1	2	1	3	2	3
8	1	3	2	3	2	1	3	1
9	1	3	3	1	3	2	1	2
10	2	1	1	3	3	2	2	1
11	2	1	2	1	1	3	3	2
12	2	1	3	2	2	1	1	3
13	2	2	1	2	3	1	3	2
14	2	2	2	3	1	2	1	3
15	2	2	3	1	2	3	2	1
16	2	3	1	3	2	3	1	2
17	2	3	2	1	3	1	2	3
18	2	3	3	2	1	2	3	1

FIGURE 7.8 Design matrix for L_{18} experiment

Standard order	COLUMN												
	1	2	3	4	5	6	7	8	9	10	11	12	13
1	1	1	1	1	1	1	1	1	1	1	1	1	1
2	1	1	1	1	2	2	2	2	2	2	2	2	2
3	1	1	1	1	3	3	3	3	3	3	3	3	3
4	1	2	2	2	1	1	1	2	2	2	3	3	3
5	1	2	2	2	2	2	2	3	3	3	1	1	1
6	1	2	2	2	3	3	3	1	1	1	2	2	2
7	1	3	3	3	1	1	1	3	3	3	2	2	2
8	1	3	3	3	2	2	2	1	1	1	3	3	3
9	1	3	3	3	3	3	3	2	2	2	1	1	1
10	2	1	2	3	1	2	3	1	2	3	1	2	3
11	2	1	2	3	2	3	1	2	3	1	2	3	1
12	2	1	2	3	3	1	2	3	1	2	3	1	2
13	2	2	3	1	1	2	3	2	3	1	3	1	2
14	2	2	3	1	2	3	1	3	1	2	1	2	3
15	2	2	3	1	3	1	2	1	2	3	2	3	1
16	2	3	1	2	1	2	3	3	1	2	2	3	1
17	2	3	1	2	2	3	1	1	2	3	3	1	2
18	2	3	1	2	3	1	2	2	3	1	1	2	3
19	3	1	3	2	1	3	2	1	3	2	1	3	2
20	3	1	3	2	2	1	3	2	1	3	2	1	3
21	3	1	3	2	3	2	1	3	2	1	3	2	1
22	3	2	1	3	1	3	2	2	1	3	3	2	1
23	3	2	1	3	2	1	3	3	2	1	1	3	2
24	3	2	1	3	3	2	1	1	3	2	2	1	3
25	3	3	2	1	1	3	2	3	2	1	2	1	3
26	3	3	2	1	2	1	3	1	3	2	3	2	1
27	3	3	2	1	3	2	1	2	1	3	1	3	2

FIGURE 7.9 Design matrix for L_{27} experiment

sponse tables was suggested to us by Randall F. Culp, Manager of Advanced Technology and Reliability Products at General Electric, Medical Products Division.

Example 7.2

A quality improvement team at Canadian Fram Limited analyzed the effects of three factors (weld time, pressure, and moisture) on the strength of a weld on an automotive part. Each factor was run at three different levels, as indicated in Figure 7.13. An L_9 design, replicated six times, was used for the experiment. This involved $9 \times 6 = 54$ experimental runs. The results of the study were reported in Carrothers (1985). The sample data and calculated values for $\bar{y}$ and *log s* are given in Figure 7.14. Response tables for $\bar{y}$ and *log s* are given in Figures 7.15 and 7.16.

The range of average responses at the bottom of Figure 7.15, over the three levels of each experimental factor, are:

- for weld time, range $= 136.1 - 123.0 = 13.1$,
- for pressure, range $= 130.5 - 121.7 = 8.8$,

Random Order Trial Number	Standard Order Trial Number	Response Observed Values y	1	2	3	1	2	3	1	2	3	1	2	3
	1													
	2													
	3													
	4													
	5													
	6													
	7													
	8													
	9													
TOTAL														
NUMBER OF VALUES														
AVERAGE														

FIGURE 7.10 Response table for L_9 experimental design

FIGURE 7.11 Response table for L_{18} experimental design

Random Order Trial Number	Standard Order Trial Number	Response Observed Values y	1:			2:			3:			4:		
			1	2	3	1	2	3	1	2	3	1	2	3
	1													
	2													
	3													
	4													
	5													
	6													
	7													
	8													
	9													
	10													
	11													
	12													
	13													
	14													
	15													
	16													
	17													
	18													
	19													
	20													
	21													
	22													
	23													
	24													
	25													
	26													
	27													
TOTAL														
NUMBER OF VALUES														
AVERAGE														

FIGURE 7.12a Response table for L_{27} experimental design

Random Order Trial Number	Standard Order Trial Number	Response Observed Values y	5:			6:			7:			8:		
			1	2	3	1	2	3	1	2	3	1	2	3
	1													
	2													
	3													
	4													
	5													
	6													
	7													
	8													
	9													
	10													
	11													
	12													
	13													
	14													
	15													
	16													
	17													
	18													
	19													
	20													
	21													
	22													
	23													
	24													
	25													
	26													
	27													
TOTAL														
NUMBER OF VALUES														
AVERAGE														

FIGURE 7.12b Response table for L_{27} experimental design (continued)

FIGURE 7.12c Response table for L_{27} experimental design (continued)

Factor	LEVELS		
	1	2	3
1. Weld time (sec)	0.2	0.4	0.6
2. Pressure (psig)	16	20	24
3. Moisture (%)	0.787	1.150	1.758

FIGURE 7.13 Levels for experimental factors

Standard order	FACTOR 1 2 3	Observed response values	$\bar{y}$	s	Log s
1	1 1 1	167.0 150.7 150.7 150.7 122.7 26.6	128.1	51.72	1.714
2	1 2 2	153.0 157.7 146.0 164.6 169.3 164.6	159.2	8.66	0.938
3	1 3 3	90.2 87.8 90.2 47.5 104.1 71.3	81.9	19.80	1.297
4	2 1 2	150.7 169.3 155.3 148.3 160.0 157.7	156.9	7.47	0.873
5	2 2 3	97.1 78.4 61.8 30.4 52.3 68.9	64.8	22.78	1.358
6	2 3 1	173.9 164.6 129.7 132.1 164.6 129.7	149.1	20.67	1.315
7	3 1 3	47.5 76.0 80.8 129.7 78.4 68.9	80.2	27.09	1.433
8	3 2 1	125.1 209.5 219.0 150.7 132.1 169.3	167.6	39.38	1.595
9	3 3 2	155.3 216.6 150.7 136.7 155.3 148.3	160.5	28.33	1.452

FIGURE 7.14 Experimental data for example

Random Order Trial Number	Standard Order Trial Number	Response Observed Values $\bar{y}$	1: Weld time 0.2 / 1	0.4 / 2	0.6 / 3	2: Pressure 16 / 1	20 / 2	24 / 3	3: moisture 0.787 / 1	1.150 / 2	1.758 / 3	4: Unassigned 1	2	3
	1	128.1	128.1			128.1			128.1			128.1		
	2	159.2	159.2				159.2			159.2			159.2	
	3	81.8	81.8					81.8			81.8			81.8
	4	156.9		156.9		156.9				156.9				156.9
	5	64.8		64.8			64.8				64.8	64.8		
	6	149.1		149.1				149.1	149.1				149.1	
	7	80.2			80.2	80.2					80.2		80.2	
	8	167.6			167.6	167.6			167.6					167.6
	9	160.5			160.5		160.5		160.5			160.5		
TOTAL		1148.2	369.1	370.8	408.3	365.2	391.6	391.4	444.8	476.6	226.8	353.4	388.5	406.3
NUMBER OF VALUES		9	3	3	3	3	3	3	3	3	3	3	3	3
AVERAGE		127.6	123.0	123.6	136.1	121.7	130.5	130.5	148.3	158.9	75.6	117.8	129.5	135.4

FIGURE 7.15 Response table for average responses

Random Order Trial Number	Standard Order Trial Number	Response Observed Values y / log s	1: Weld time (0.2 / 0.4 / 0.6)			2: Pressure (16 / 20 / 24)			3: moisture (0.787 / 1.150 / 1.758)			4: Unassigned		
			1	2	3	1	2	3	1	2	3	1	2	3
	1	1.719	1.714			1.714			1.714			1.714		
	2	0.938	0.938				0.938			0.738			0.938	
	3	1.297	1.297					1.297			1.297			1.297
	4	0.873		0.873		0.873				0.873				0.873
	5	1.358		1.358			1.358				1.358	1.358		
	6	1.315		1.315				1.315	1.315				1.315	
	7	1.433			1.433	1.433					1.433		1.433	
	8	1.595			1.595		1.595		1.595					1.595
	9	1.452			1.452			1.452		1.452		1.452		
TOTAL		11.975	3.949	3.546	4.480	4.020	3.891	4.064	4.629	3.263	4.088	4.524	3.686	3.765
NUMBER OF VALUES		9	3	3	3	3	3	3	3	3	3	3	3	3
AVERAGE		1.331	1.316	1.182	1.493	1.340	1.297	1.355	1.541	1.088	1.363	1.508	1.229	1.255

FIGURE 7.16 Response table for log s values

- for moisture, range $= 158.9 - 75.6 = 83.3$, and
- for unassigned "factor 4," range $= 135.4 - 117.8 = 17.6$.

The effect of unassigned factor 4 reflects random variability. The ranges of average weld strength as weld time or pressure change are less than the range attributed to random variability. We can conclude, then, that changes in weld time or pressure do not have a significant effect on the average weld strength. Moisture does appear to affect weld strength, however. As can be seen in Figure 7.17, weld strength is much lower when moisture is 1.758 percent. The other two levels would be preferred, since the object is to maximize weld strength.

If the ranges of average values for *log s* at the bottom of Figure 7.16 are calculated over the three levels of each experimental factor, we obtain:

- for weld time, range $= 1.493 - 1.182 = 0.311$,
- for pressure, range $= 1.355 - 1.297 = 0.058$,
- for moisture, range $= 1.541 - 1.088 = 0.453$, and
- for unassigned "factor 4," range $= 1.508 - 1.229 = 0.279$.

Again, only moisture seems to have any real effect. *Log s* is minimized when moisture is 1.150 percent. This is consistent with the results based on average

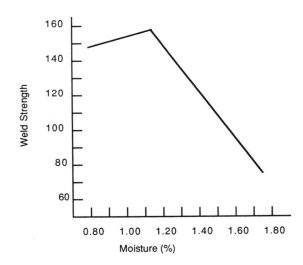

FIGURE 7.17 Weld strength as a function of moisture

weld strength discussed above. The recommendation from the study, then, should be to set moisture at about 1.150 percent.

Before leaving this example, a caveat should be added. The response measurements in Figure 7.14 showed large variability, but there were also a number of measurements which were equal to each other. It is not clear from the original report whether the trials were randomized, but there may be some distorting forces at work.

8

Analysis of Variance in Engineering Design

8.1 Hypothesis Testing Concepts

Julia, a manufacturing engineer at an electrical machinery company, developed a new process for manufacturing motor housings. She claimed that the new approach would increase the production rate without affecting product characteristics. A review committee agreed that her idea had merit, and decided to try the new approach on a pilot basis. Because of the costs related to changing production processes, the committee decided that they would recommend the new process if the pilot study showed an increase in production rate of at least 8 percent. One assembly area was designated for the pilot study. Data were gathered for three weeks on the number of motor housings produced per hour under the current process. Then manufacturing equipment was modified as needed and workers were trained in the new method. The new process was then implemented. After two weeks of production, data were gathered on production rates under the new procedure for the following three weeks. It was found that under the new process production rates increased by 11.2 percent. The review committee recommended that the new process be implemented for all relevant production lines.

In the above example, the review committee performed a statistical hypothesis test, which involved:

1. Formulating two *hypotheses* or conjectures about what was going on. The committee's two hypotheses were (i) the new process has a higher production rate than the old process and (ii) the new process does not have a higher production rate.
2. Selecting a *test statistic* and *critical value* for that test statistic. A *test statistic* is a number calculated from sample data which is used as the basis for deciding which hypothesis appears to be the correct one. A *critical value* is a number, agreed upon before the data are gathered, such that if the test statistic is greater than that number, one hypothesis will be se-

lected as the "correct" one, while if the test statistic is less than that number, the other hypothesis is selected. (Which hypothesis is chosen if the test statistic is equal to the critical value depends on the situation.) For the example we are considering, the test statistic is the average hourly production rate under the new process minus the average hourly production rate under the old production process, based on a total of six weeks of data. The critical value is 8 percent.

3. Gathering data to be used to calculate the test statistic value. The committee observed, over two three-week periods, the production rates for motor housings using the old and new processes.

4. Determining the significance of the data. That is, they calculated the value of the test statistic and decided which hypothesis was supported by the data, based on the pre-established critical value. For this example the test statistic was equal to 11.2 percent. Since this is larger than the critical value of 8 percent, the committee accepted the hypothesis that the new manufacturing process really was better.

Steps 1 through 4 above illustrate the basic elements of a classical hypothesis test. Note that the two hypotheses proposed in step 1 were not treated equally. If the new process had increased production rates by, say, 5 percent, the hypothesis of no improvement with the new process would have been chosen, even though there was some evidence of improvement. That is, the review committee wanted evidence that the new process was much better than the old before they would consider recommending the new method. In a traditional test of hypothesis, the hypothesis which is "accepted" when the data are not conclusive is called the *null hypothesis*. The other hypothesis, the one which is accepted only if the data show strong evidence that it is the correct one, is called the *alternative hypothesis*. For this example, the null hypothesis is that the new process does not have a higher average production rate than the old process and the alternative hypothesis is that the new process does have a higher production rate.

Hypothesis testing is much like a criminal trial. In a trial there are two hypotheses:

the accused is not guilty, and

the accused is guilty.

Data (testimony and evidence) are gathered, and the jury evaluates the evidence. At the end of the trial, one of the hypotheses is selected as being the correct one, based on the data. If the data are inconclusive the null hypothesis of "not guilty" is selected. The "burden of proof" is always on the alternative hypothesis (accused is guilty).

In the above discussions of hypothesis testing, there was no mention of the data "proving" which hypothesis is correct. In traditional hypothesis testing it is usually not known for sure which hypothesis is correct—the data can mislead us. Two types of errors are possible:

Type I error: Accept a false alternative hypothesis, and

Type II error: Accept a false null hypothesis.

A well-designed experiment will keep the likelihood of both types of error at an acceptable level. This can be difficult, since in general if you try to decrease the likelihood of one type of error occurring, you automatically increase the likelihood of the other error type occurring.

In this chapter we will see how to perform hypothesis tests for factor effects (including interactions). The null hypothesis will be that the factor under consideration has no significant effect on the response variable. The alternative hypothesis will be that the factor has a significant effect on the average response. Here *significant* effect means that it is unlikely that the observed estimated effect could be as large as it is just by chance—that there is strong evidence of a real factor effect. Test statistics will be calculated from the estimated effects and critical values will be obtained from a table of values developed for that purpose.

Some of the material in this chapter, particularly section 8.2, is more advanced than is found elsewhere in the book. This may be a problem for readers who have not been exposed previously to statistics. But the methods presented here can be applied without understanding completely the material in section 8.2 and, as will be discussed at the beginning of section 8.2 and in section 8.4, normal plots are often more useful than analysis of variance when analyzing data from engineering experiments.

8.2 Using Estimated Effects as Test Statistics

In section 3.6 it was recommended that all estimated effects from an experiment be plotted on normal probability paper. If there are no "real" factor or interaction effects, then the estimated effects should have a distribution similar to that of a random sample from a normal population. When plotted on normal probability paper, these estimated effects should be in approximately a straight line. Effects which *are* real, on the other hand, should have estimates which appear too large or too small to be part of a normal random sample. On normal probability paper, these estimates will be outside a straight-line pattern displayed by the bulk of the other points, and will be classified as outliers. This graphical approach to identifying "significant" estimates is useful because it provides some protection against identifying an effect as being real simply because it is larger, but not significantly larger, in magnitude than other estimated effects; all points will still follow a straight-line pattern. The graphical approach has the disadvantage of not providing a clear criterion for what values for estimated effects indicate *significant* factor or interaction effects. That is, how far must a point be from the straight-line pattern before it is judged to be an outlier? And how do we measure amount of departure from the straight line pattern? *Analysis of variance* (ANOVA) meets this need by specifying by how much an estimate must differ from zero in order to be judged "statistically significant." Note that we said "how much an estimate must differ from *zero*," not "how much it must depart from the *straight-line pattern*." A serious weakness with traditional use of analysis of variance is that the test statistics do not take into account the increased variability to be expected when many effects are being estimated. When, say, fifteen or more effects are estimated and individually tested in the same experiment, there is a good chance that a classical test of hypothesis will detect as "significant" an effect which is further from zero than the rest, but does not depart from the overall straight-line pattern. This is the type I error mentioned in section 8.1. (An area of statistics called "multiple comparisons" does address this problem; see Box, Hunter, and Hunter, 1978, or Ott, 1984, for further discussion on this topic.) So although analysis of variance is a more quantitative alternative to normal plotting, it is not necessarily a better technique. The advantages and disadvantages of using analysis of variance will be explored further in section 8.4.

Analysis of variance is based on the following facts, which will be stated here without proof. (Proofs can be found in most texts on mathematical statistics; see Box, Hunter, and Hunter, 1978.)

Fact 1: If a given factor or interaction from a two-level experiment has no real effect on the response variable, and if the individual measurements follow a normal distribution with constant standard deviation, then the estimated effect for this factor or interaction will have a normal probability distribution with mean *0* and standard deviation $2\sigma/\sqrt{n}$, where σ is the standard deviation of the individual response measurements obtained in the experiment and n is the number of measurements obtained. (Note that we are assuming here that the standard deviation of response values is constant throughout the experiment—a blatantly dangerous assumption, as was pointed out in section 5.1.)

Fact 2: Suppose $E_1, E_2, E_3, \ldots, E_j$, and E_i are $j+1$ orthogonal estimates of factor effects or interaction effects from a two-level experiment. That is, $E_1, E_2, \ldots, E_j$, and E_i are the estimates calculated at the bottom of a response table, or some subset of those estimates. Suppose further that the conditions given in Fact 1 apply and the effects being estimated by $E_1, E_2, \ldots, E_j$, and E_i are not real (that is, have expected value of zero). Then:

$$F = E_i^2/[(E_1^2 + E_2^2 + \ldots + E_j^2)/j]$$

will follow an *F distribution* with *(1, j) degrees of freedom*.

In section 5.3 the normal probability distribution was described. The F distribution is another probability distribution with a different shape. The actual form of the distribution of F is not important here. What is important is that the probability distribution of this statistic can be calculated, and so we can tell when particularly unusual events have occurred. This situation is somewhat analogous to knowing the probability distribution of poker hands. In order to win at poker it is helpful to know the probability of being dealt a pair or a straight, but the mathematics involved in calculating those probabilities is not important to the gambler. "Degrees of freedom" refer to a parameter which identifies the particular form of F distribution—the shape of the F distribution changes as the number of estimated effects in the denominator changes.

Fact 2 tells us what to expect from F if the Es are all estimating small or nonexistent effects. But if E_i is estimating a real effect, then we do not expect E_i to be close to zero, and so we expect to see F larger than it would have been had E_i been estimating a small or nonexistent effect. So we can use F as a test statistic to test the null hypothesis—

H_N: E_i estimates a nonexistent effect or interaction

against the alternative hypothesis—

H_A: E_i estimates a real effect or interaction.

If F is "close" to one, we decide we have insufficient reason to reject H_N. If F is larger than some number we obtain from an F distribution table, we reject H_N in favor of H_A. The "critical value" we pick from the F table has the property that the probability of F being larger than that number is known when there is no real effect. Table 8.1 gives critical F values for which the probability of an F statistic exceeding the table values is 1 percent and 5 percent if the conditions described in Fact 2 apply (and so H_N is true). The table is appropriate for use with effects estimated from two-level experiments. For factors at three or more levels, different formulas would be needed to calculate the F statistic value and the first degree of freedom for the F statistic would be greater than one. Most statistics texts have tables of the F distribution for the first degree of freedom greater than one.

Example

The concepts in the above paragraph can be best understood through an example. Consider the example from section 4.1 in which an engineer sought to maximize the bond strength when mounting an integrated circuit (I.C.) on a metalized glass substrate. Four factors were identified as possibly affecting the bond strength. They were listed in Figure 4.5, which is reproduced as Figure 8.1. The factor levels used in the experiment are also given in Figure 8.1. The estimated effects were calculated in a response table in Figure 4.7. The estimates, and their squared values, are given in Figure 8.2. If it can be assumed that there are no interaction effects among the four experimental factors, then the average of the squares of the estimated interaction effects (columns 4, 5, and 6 of the response table) can be used as the denominator for the F statistic described in "Fact 2." The value of j in Fact 2 will then be equal to 3. The information from Figure 8.2 is used to construct the analysis of variance, or "ANOVA," table in Figure 8.3. The first four E^2 values in Figure 8.3 come directly from Figure 8.2. The E^2 value in the "no effect" row is the average of the three E^2 values for interaction effects in Figure 8.2. That

	SIGNIFICANCE LEVEL	
j	*5%*	*1%*
1	161.4	4052.
2	18.51	98.50
3	10.13	34.12
4	7.71	21.20
5	6.61	16.26
6	5.99	13.74
7	5.59	12.25
8	5.32	11.26
9	5.12	10.56
10	4.96	10.04
11	4.84	9.65
12	4.75	9.33
13	4.67	9.07
14	4.60	8.86
15	4.54	8.68
16	4.49	8.53
17	4.45	8.40
18	4.41	8.29
19	4.38	8.18
20	4.35	8.10
21	4.32	8.02
22	4.30	7.95
23	4.28	7.88
24	4.26	7.82
25	4.24	7.77
26	4.23	7.72
27	4.21	7.68
28	4.20	7.64
29	4.18	7.60
30	4.17	7.56
40	4.08	7.31
60	4.00	7.08
120	3.92	6.85
∞	3.84	6.63

TABLE 8.1 F distribution critical values for $(1, j)$ degrees of freedom

Factor	*Low level*	*High level*
A. Adhesive type	D2A	H-1-E
B. Conductor material	copper	nickel
C. Cure time (at 90° C)	90 min	120 min
D. I.C. post coating	tin	silver

FIGURE 8.1 Factor levels for example experiment

Response table column	Factors	Effect estimate, E	Estimate squared, E²
1	A: Adhesive type	2.2	4.84
2	B: Conductor material	−0.8	0.64
3	C: Cure time	6.2	38.44
4	AB & CD interactions	0.8	0.64
5	AC & BD interactions	0.8	0.64
6	AD & BC interactions	−0.2	0.04
7	D: I.C. post coating	9.2	84.64

FIGURE 8.2 Estimated factor and interaction effects

is, the "no effect" $E^2 = (0.64 + 0.64 + 0.04)/3 = 1.32/3 = 0.44$. "df" in Figure 8.3 refers to "degrees of freedom," which in this case is just the number of factors used in the row. "F" refers to F statistics, with $(1,3)$ degrees of freedom, which are calculated by dividing the E^2 value in a row by the *no effect* E^2. The calculated F values should be compared to theoretical extreme values for the F distribution. From Table 8.1 the upper 5 percent value for an F distribution with $(1,3)$ degrees of freedom is 10.13; the upper 1 percent value is 34.12. That is, if a factor has no real effect on the response variable, then the probability that its calculated F statistic will be larger than 10.13 is one in twenty, or 5 percent. The probability that its F statistic will exceed 34.12 is only one in a hundred (1 percent). Comparing the calculated F values from Figure 8.3 to the tabulated F values from Table 8.1, we can conclude that:

- factor A affects the average response at the 5 percent level of statistical significance;
- factor B does not have a significant effect on the average response value;
- factors C and D affect the average response at the 1 percent level of statistical significance.

So factors C and D have the greatest impact on bond strength. Factor A has a lesser effect. Factor B appears to have no significant effect. These conclusions are consistent with those reached when these data were analyzed graphically in section 4.1.

Factor	E²	df	F
A: Adhesive type	4.84	1	11.00
B: Conductor material	0.64	1	1.45
C: Cure time	38.44	1	87.36
D: I.C. post coating	84.64	1	192.36
No effect	0.44	3	

FIGURE 8.3 Analysis of variance table for example

8.3 Analysis of Variance for Two-level Designs

In the example at the end of section 8.2, an analysis of variance was performed on a two-level design. This technique can be used with any two-level full factorial or fractional factorial design.

Steps in analysis of variance for unreplicated two-level designs

1. Carry out a two-level experiment, as explained in Chapters 3 and 4. Calculate estimated effects using a response table.

2. Determine which columns in the response table do not represent real effects. Square their estimated effects and average these squares. This average will be the "no effect" E^2. In deciding which columns might estimate real effects, engineering knowledge and previous experience should be used. It is not appropriate to base this decision on which estimates are small, since random variability can produce both large and small estimates. If effects are selected for inclusion in the "no effect" category based only on values of estimates, the result will often be a denominator for the F statistic which is too small, leading to artificially large F statistics. Rules which say "pool the 50 percent of the estimated effects which are closest to zero" or "pool the one-third of the estimated effects which are closest to zero" will often get the experimenter into trouble.

3. Prepare an ANOVA table similar to Figure 8.3:

- in the first column put factor and interaction names. Include "no effect" at the bottom of the column.
- in the second column put the squares of the estimated effects. For the "no effect" row enter the average of the squared "no effect" effects, as explained earlier.
- put "1" in the df ("degrees of freedom") column for each estimated effect. For the "no effect" row enter the *number* of estimators included in calculating the no effect E^2.
- for each effect listed in the ANOVA table, divide the square of the estimated effect by the no effect E^2 value. Compare this number to F table values, using (1,j) degrees of freedom, where j is the no effect degrees of freedom. Calculated F values larger than table F values suggest that the factor is having an influence on the response variable.

Some readers may be familiar with the usual approach of preparing ANOVA tables by first calculating a sum of squares value for each row of the table, then calculating a mean square value for each factor by dividing the sum of squares by degrees of freedom, and finally calculating F values as ratios of mean square values. The approach here produces the same answers, but is easier to do. Traditional mean square values can be obtained by multiplying E^2 values by the number of experimental trials and dividing this product by four. The "no effect" E^2 value is an estimate of the variance for the E values. The "no effect" mean square value estimates the variance for individual response values. Introductory discussion on classical analysis of variance can be found in Box, Hunter, and Hunter (1978), Ott (1984), and many other textbooks on statistics and experimental design.

The approach to calculating F values suggested in this chapter is not new, and works for orthogonal two-level experimental designs. See Guttman, Wilks, and Hunter (1971) for further discussion on this method. For designs with factors at three or more levels the traditional approach to calculating an ANOVA table must be used. We will not consider the calculation of such tables in this chapter, but in Chapter 9 we will consider statistical software which can perform these calculations for us.

Example

The last example in section 4.3 involved analyzing the effects of twelve factors in sixteen runs. The factors and their experimental levels are given in Figure 4.34. The completed response table from the experiment is given in Figure 4.35. The estimated effects obtained in Figure 4.35 are also given in the third column in Figure 8.4. The squares of these estimates are given in the last column of Figure 8.4. If it can be assumed that there are no interactions among the experimental factors, then the squares of the estimates from columns 8, 9, and 10 from Figure 8.4 can be averaged to obtain the no effect E^2 value. The ANOVA table for these data is given in Figure 8.5. The values in the last column of Figure 8.5 were obtained by dividing each squared effect by the no effect squared effect. From Table 8.1, the 5 percent and 1 percent F values for (1,3) degrees of freedom are 10.13 and 34.12, respectively. Comparing the F values in Figure 8.5 with these 1 percent and 5 percent values for the F

Column	Factor	Estimated effect, E	E^2
1	A	5.9	34.81
2	B	10.4	108.16
3	C	6.1	37.21
4	D	−0.4	0.16
5	J	0.6	0.36
6	K	1.4	1.96
7	L	15.9	252.81
8	—	0.9	0.81 ⎤
9	—	2.9	8.41 ⎬ 3.08
10	—	0.1	0.01 ⎦
11	E	−12.9	166.41
12	F	8.1	65.61
13	G	−0.6	0.36
14	H	1.4	1.96
15	I	8.6	73.96

FIGURE 8.4 Estimated effects and their squares

Column	Factor	E^2	df	F
1	A	34.81	1	11.3
2	B	108.16	1	35.1
3	C	37.21	1	12.1
4	D	0.16	1	0.1
5	J	0.36	1	0.1
6	K	1.96	1	0.6
7	L	252.81	1	82.1
11	E	166.41	1	54.0
12	F	65.61	1	21.3
13	G	0.36	1	0.1
14	H	1.96	1	0.6
15	I	73.96	1	24.0
8–10	No effect	3.08	3	

FIGURE 8.5 ANOVA table for example

probability distribution, we find that factors B, E, and L are significant at the 1 percent level, and factors A, C, F, and I are significant at the 5 percent level. These are the same seven factors which were identified as most important when the data were analyzed graphically in section 4.3.

Analysis of variance with blocked experiments

If an experiment is blocked (see section 4.5), the columns which are assigned to block effects should not be used as part of the no effect E^2 term. Also, an F statistic is not computed for block effects since block effects are not generally of interest to the experimenter. Other than that, the handling of blocked designs is the same as for unblocked designs.

Analysis of variance with replicated experiments

For replicated experiments, the no effect E^2 is based on the sample variances calculated for each "point" in the experimental design. The sample variances are squares of sample standard deviations. The procedure for calculating them is explained in section 5.2. The formula for calculating the no effect E^2 based on these individual sample variances is:

$$E^2 = 4s^2/N$$

where s^2 is the average of the sample variances calculated at the experimental "points" and N is the total number of trials performed (number of different experimental points times the number of replications). The degrees of freedom for the no effect E^2 is equal to N minus the number of different experimental points.

Example

The example at the end of section 5.4 involved an eight-run fractional factorial design, replicated five times. So $N = 8 \times 5 = 40$. The no effect degrees of freedom will be $40 - 8 = 32$. The raw data and averages and sample standard deviations at each of the eight experimental points are given in Figure 5.13. The estimated effects on the response average are calculated in the response table in Figure 5.14. Based on the standard deviations in Figure 5.13, the pooled sample variance for this experiment would be:

$$s^2 = \frac{1.57^2 + 0.71^2 + 0.80^2 + 1.41^2 + 2.06^2 + 2.93^2 + 5.17^2 + 1.99^2}{8}$$

$$= \frac{2.46 + 0.50 + 0.64 + 1.99 + 4.24 + 8.58 + 26.73 + 3.96}{8}$$

$$= 49.10/8$$

$$= 6.14,$$

and the no effect E^2 would then be:

$$E^2 = 4s^2/N$$
$$= 4 \times 6.14/40$$
$$= 0.614.$$

Based on the response table in Figure 5.14, the ANOVA table for the main effects of this experiment is given in Figure 8.6.

Table 8.1 does not list critical F values for (1,32) degrees of freedom. However, if we interpolate between the F values for (1,30) and (1,40) degrees of freedom taken from Table 8.1, that the upper 5 and 1 percent values for the F distribution with (1,32) degrees of freedom are approximately 4.15 and 7.51, respectively. Factor A has an F value between 4.15 and 7.51. Factors C and D have F values larger than these two numbers. We conclude that factors C and D have a highly significant effect on the average response, factor A has a less significant effect, and factor B has no statistically significant effect on average response.

Analysis of variance using center points

It sometimes happens that no column of a response table can be treated as a "no effect" factor, and it is impractical or too expensive to replicate the experiment. In these situations it may be possible to identify a "center point," a point where *all* factor levels are halfway between their low and high values.

Factor	E	E²	df	F
A: Adhesive	1.96	3.84	1	6.25
B: Conductor	0.73	0.53	1	0.86
C: Cure time	5.44	29.59	1	48.19
D: Post coat	8.71	75.86	1	123.55
No effect		0.614	32	

FIGURE 8.6 ANOVA table for replicated experiment

If this point can be replicated a few times, a sample variance can be calculated at this point using the formula given in section 5.2. Assuming constant variance at all levels of the experimental factors (this will be discussed further in the next section), the "no effect" E^2 value can then be calculated by multiplying this sample variance by 4/N, where N is the number of trials in the experiment, *excluding the center points.* The "no effect" degrees of freedom is equal to the number of center points minus one. Another use of the center point is to compare the average response at the center point with the average at the other points. If these two averages differ greatly, this suggests a nonlinear relationship between average response and some of the experimental factors.

This whole idea of a center point assumes that each design factor has a "middle value." If one factor was Production Line, and a plant only had two production lines, then a center point using "Production Line 1.5" doesn't make sense. In this case, either we shouldn't use a center point, or we must assume that response variability is the same on both production lines, and arbitrarily assign one of the production lines to the center point.

8.4 When to Use Analysis of Variance

Analysis of variance has been used for many years by engineers, statisticians, and a variety of other experimenters. It can be applied not only to two-level experiments, but also to three-level, four-level, and many other experimental designs. Analysis of variance can also be used with experiments having *log s* or the S/N ratio as the response variable. How analysis of variance is applied to these other situations will not be described here, however, because the authors feel, for two reasons, that the graphical methods described earlier in this book are generally preferable to analysis of variance for most engineering experiments:

1. There is a difference between *statistical* significance and *practical* significance. Whether or not an estimated factor effect is statistically significant is a function of the magnitude of the estimated effect, the size of the no effect E^2 and the size of the F value you obtain from Table 8.1. As no effect degrees of freedom gets larger, through replication or adding effects to the no effect "pool," the F numbers in Table 8.1 get smaller. So it can (and does) happen that a rather large estimated effect will not be statistically significant because the number of experimental trials is small. Similarly, a factor which has an insignificant effect from a practical point of view can be statistically significant because the experiment was highly replicated.

2. There are some underlying assumptions made when using analysis of variance which are often not true. In particular, it is assumed that estimated effects are random variables which have a normal distribution. It is also assumed that the *variability* of the response variable is the same at all points in the experimental design. This may never be true in engineering experiments. Dr. Taguchi has stressed this point. One of the most important uses of experimental designs is to identify what combination of factor levels minimizes response variability.

As was mentioned in section 8.2, normal plots do not provide clear criteria for deciding when a point is "significantly" far away from the pattern of the other points to be judged an outlier. However, in many engineering studies, additional data can be gathered if initial analyses are inconclusive. When planning experiments, time and resources should always be allocated for confirmatory runs and for further investigation if needed. This situation sets industrial experiments apart from most academic research. In academic studies the experimenter is often expected to state formal hypotheses, perform a classical test of hypothesis, and report results based on a predetermined test statistic. In such situations, analysis of variance may make a lot of sense. Unfortunately, people sometimes try to fit industrial experiments into traditional academic molds, and this often produces less than satisfactory results. In most engineering experiments, the experimenter is better off using normal probability plots and other methods presented in earlier chapters of this book rather than analysis of variance.

9

Computer Software For Experimental Design

9.1 Role of Computer Software in Experimental Design

All of the experimental designs presented in this book can be analyzed without use of a computer. But statistical packages can simplify the task of the experimenter and reduce computational errors. Some software packages will provide the experimental design you need, do blocking, print out a sheet for recording the results of the experiment, calculate the estimated response values, and provide normal plots and analysis of variance tables. Why then even bother with the hand calculations we went through in earlier chapters? Data analysis should never become a mechanical process where we blindly accept numbers spit out by a computer algorithm. If unexpected or unusual estimates appear, we should do some investigating. What caused this estimated effect to be so large? Is there one response value which is way out of line? Is there a lot of variability in the data? Were our choices for design factor levels inappropriate? If we don't know how estimates are calculated, we can't answer these questions. But having gone through the steps of hand calculation, we are now in a position to use experimental design software *intelligently*.

The methods presented in this book for two-level experiments in eight and sixteen runs are readily generalizable to two-level experiments in thirty-two, sixty-four, 128, or any number of runs which is a power of two. There is also a class of two-level designs developed by Plackett and Burman (1946) which require that the number of trials be equal to a multiple of four, instead of a power of two. Plackett-Burman designs have complicated confoundings of interaction effects with each other and with main effects, and so are of use primarily in screening experiments where a large number of factors are being considered and interactions are being ignored. Plackett-Burman designs allow analysis of up to n-1 factors in n trials. If Plackett-Burman designs are folded over (see sections 4.2 and 4.4), the resulting design will be of resolution IV, which means that no main effects estimates will be confounded with two-factor interaction effects. Based on what you have learned in earlier chapters, you should have no problem dealing with these larger experiments, with the aid of a good software package.

Experimental designs for factors at three or four levels were presented in Chapter 7. These experiments can also be analyzed using statistical software. The more sophisticated software packages ask the experimenter how many factors are to be used in the experiment and what the number of levels is for each factor, and then presents an appropriate experimental design. Such packages will also calculate estimated values of average responses and of standard deviations and an analysis of variance table. A word of caution should be added here. It is tempting to run each factor at three, four, or five levels and let the computer do the calculations. But two-level designs are not something you graduate out of—they are the most valuable of all experimental designs for engineering application. With higher numbers of factor levels, the number of experimental runs increases dramatically, interaction effects are difficult to detect, and analysis of variance replaces normal plots as the basic tool for measuring significance of effects. An alternate approach to using more than two levels per factor is to view engineering experiments as an ongoing process rather than as a one-shot deal. That is, run a fractional factorial experiment with all factors at two levels, analyze the results, design another experiment, and repeat the cycle. Continue until the key factors affecting the response are known and their interactions understood.

The situation described above could also be attacked using *multiple regression*, a statistical approach in which a model is constructed to explain the behavior of a response variable as a function of several experimental factors. Analysis of variance and graphical techniques are used to evaluate the adequacy of the proposed model and the relative importance of the experimental factors. (See Ott, 1984, for a basic introduction to multiple regression, or Draper and Smith, 1981, for a more thorough discussion.) Another possibility is to use *response surface methodology*, in which the response variable is treated as a function of continuous experimental factors, and the relationship between the factors and the response is evaluated with the intent of maximizing or minimizing some function of the response variable (see Box and Draper, 1987). The reader is strongly encouraged to learn more about multiple regression and response surface methodology. They are valuable tools for analyzing engineering data and for studying relationships among variables. Some software packages include response surface methods and can generate contour plots and three-dimensional plots of the response surface.

9.2 Summary of Statistical Packages

It is impossible to write a "current" summary of existing software—the field doesn't stand still long enough. Christopher Nachtsheim (1987) provided an excellent review of eleven packages useful in experimental design. But new products have entered the market since then, and existing packages have been improved. In this chapter we will give examples of experimental design and data analysis using the five software packages listed below.* They represent a fairly broad range of sophistication and price. Their operating environments are listed in Figure 9.1. Their capabilities with respect to the material covered in this book are summarized in Figure 9.2.

DESIGN-EASE

Jass

NCSS Statistical System

RS/Discover Software

SCA Statistical System

When purchasing software, it is important to consider future data analysis needs as well as current requirements and price. A useful reference when shopping for software for statistical and quality applications is the March issue of *Quality Progress*, published by the American Society for Quality Control. This issue contains an annual "QA/QC Software Directory."

Software package	Computer	Operating system	Minimum memory
DESIGN-EASE	IBM PC	MS-DOS	256K
Jass (Version 2.1)	IBM PC	MS-DOS	512K
	VAX	VMS, UNIX	
NCSS Statistical System (Version 5.02)	IBM PC	MS-DOS	512K
RS/Discover Software (Release 2.01)	VAX	VMS, ULTRIX Berkley UNIX	
	MicroVAX	MicroVMS	
	SUN 3,4	SUN UNIX	
	HP 300, 800	HP-UX	
	DECSTATION	ULTRIX	
	VAXSTATION	ULTRIX	
SCA Statistical System (Version IV.I)	IBM PC	MS-DOS	640K
	IBM mainfrm	MVS, CMS, MTS	
	VAX	VMS, UNIX	
	PYRAMID	UNIX	
	UNIVAC	EXEC	
	CDC	NOS	
	HP 9000	HP-UX	
	APOLLO		
	MicroVAX	VMS	
	SUN	UNIX	

FIGURE 9.1 Hardware and operating systems for software

*For information on the publishers of these software packages, see page 243.

	DESIGN EASE	Jass	NCSS	RS/Discover	SCA
Two-level factorial design	X	X	X	X	X
with blocking	X	X	X	X	X
replicated	X	X	X	X	X
replicated, with blocking	X	X	X	X	X
Two-level fractional designs	X	X	X	X	X
with blocking	X	X	X	X	X
replicated		X		X	X
replicated, with blocking		X		X	X
Plackett-Burman designs	X		X	X	X
Fold-over designs				X	X
ANOVA	X		X	X	X
Normal plots	X	X	X	X	X
Interaction plots	X	X	X	X	X
3-D cube plots*	X	X	X		X
Report forms	X	X	X	X	
Menu-driven	X		X	X	
On-line help	X	X	X	X	X
Other statistical routines			X	X	X

Cube plots are explained in Example 1 of section 9.2. Figure 9.13 is an example of a cube plot.

FIGURE 9.2 Features of software packages in relation to material covered in this book

9.3 Examples of Use of Software Packages

The software packages considered in this chapter generally use the traditional Western ordering of experimental levels as discussed in section 3.10 and shown in Figures 3.64, 3.65, and 3.66. Examples of two-level experiments presented in this book can be run using Western or Taguchi software if the factors are assigned to different design matrix columns when using these packages (and, for Taguchi software, if the high and low levels for factors are interchanged). For example, the factor assignments needed to run our examples using most Western software are given in Figure 9.3. This is not a problem when analyzing actual experiments since the software will often identify the arrangement of factor levels to be run and the resulting interaction effects.

1. Full-factorial two-level design in eight runs

To see how this works, consider the mechanical plating example given in section 3.7. The three experimental factors and their levels are given in Figure 3.37. This figure is reproduced as Figure 9.4. Based on the rules in Figure 9.3 for assignment of factors to columns of the design matrix, in order to reproduce the example in section 3.7, we would list the first factor in Figure 9.4, *Cleaner*, in the third column of the design matrix for DESIGN-EASE, and the third factor in Figure 9.4, *Zinc*, in the first column. The second factor, *Copper*,

L₈ DESIGN			L₁₆ DESIGN		
THIS BOOK		*Western software name*	THIS BOOK		*Western software name*
Column	*Name*		*Column*	*Name*	
1	A	C	1	A	D
2	B	B	2	B	C
3	C	A	3	C	B
4	AB	BC	4	D	A
5	AC	AC	5	AB	CD
6	BC	AB	6	AC	BD
7	ABC	ABC	7	AD	AD
			8	BC	BC
			9	BD	AC
			10	CD	AB
			11	ABC	BCD
			12	ABD	ACD
			13	ACD	ABD
			14	BCD	ABC
			15	ABCD	ABCD

FIGURE 9.3 Design matrix factor names in this book and standard Western software

would remain in the second place. DESIGN-EASE provides the user with a printed report form of the type shown in Figure 9.5. Figure 9.5 has the same combinations of factor levels found in Figure 3.38. The random orderings of the experimental runs, as expected, are different for Figures 3.38 and 9.5.

Once the report form in Figure 9.5 has been completed, the data are entered into the computer and DESIGN-EASE provides the estimated factor effects as shown in the third column of Figure 9.6. Keeping in mind that factors A, B, and C in Figure 9.6 are Zinc, Copper, and Cleaner, respectively, the estimates in Figure 9.6 are the same as those obtained by hand calculation in Figure 3.39.

DESIGN-EASE also provides some graphical analyses of main and interaction effects, as well as normal plots. The main effect for copper is plotted in Figure 9.7. The interaction between zinc and cleaner is given in Figure 9.8. This figure can be compared with Figure 3.43-b (remember the A and C factors have been interchanged). Note on Figures 9.7 and 9.8 that the y-axis indicates the largest ("Max") and smallest ("Min") response values observed during the experiment. Other software packages provide similar graphical

Factor	*Level 1*	*Level 2*
A: Cleaner	90% of recipe	recipe
B: Copper	71% of recipe	recipe
C: Zinc	67% of recipe	recipe

FIGURE 9.4 Levels of factors for experiment

D E S I G N - E A S E   A N A L Y S I S

Run Ord	Blk	A zinc percent	B copper percent	C cleaner percent	R1 y	Std Ord
1	1	100	100	100	_____	8
2	1	67	71	90	_____	1
3	1	100	100	90	_____	4
4	1	100	71	100	_____	6
5	1	67	71	100	_____	5
6	1	67	100	100	_____	7
7	1	100	71	90	_____	2
8	1	67	100	90	_____	3

FIGURE 9.5 Report form prepared by DESIGN-EASE

displays. An interesting feature of DESIGN-EASE's normal plotting routine is that you can move a straight line around on the screen with the normal plot to obtain a visual "best fit." Figure 9.9 shows such a fit of a line through the center of the normal plot of estimated effects. The normal plot in Figure 9.9 was done by hand in Figure 3.41.

DESIGN-EASE also provides an analysis of variance table for experiments. But instead of providing F values, it supplies t values (see Figure 9.10). An F value can be obtained from a t value by squaring the t value. That is, $F = t^2$. The resulting F statistic will have $(1, v)$ degrees of freedom, where v is the degrees of freedom for the t statistic. But this conversion is usually unnecessary. The level of significance for the t value, and for the F value as well, can be read from the last column of Figure 9.10. Most software packages

Dep Variable: y in   DESIGN-EASE
ANALYSIS

VARIABLE	COEFFICIENT	EFFECT	SUM OF SQUARES
mean	57.750000	57.750000	
A	5.500000	11.000000	242.000000
B	2.500000	5.000000	50.000000
AB	0.250000	0.500000	0.500000
C	-2.750000	-5.500000	60.500000
AC	-4.000000	-8.000000	128.000000
BC	3.000000	6.000000	72.000000
ABC	0.750000	1.500000	4.500000

FIGURE 9.6 DESIGN-EASE estimated effects for example

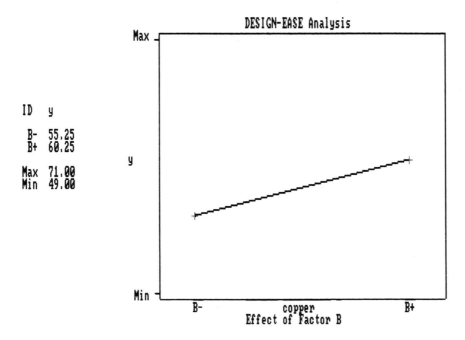

FIGURE 9.7 DESIGN-EASE plot of main effects for copper

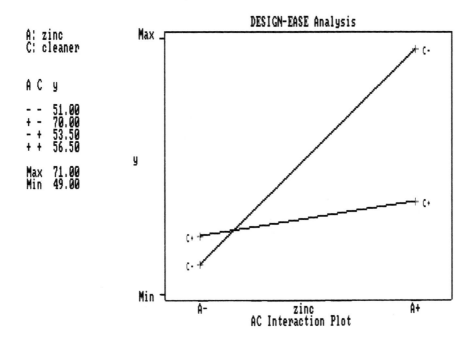

FIGURE 9.8 DESIGN-EASE plot of zinc-cleaner interaction

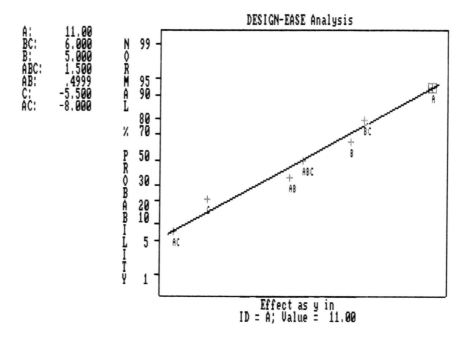

FIGURE 9.9 DESIGN-EASE normal plot of estimated effects

provide significance levels with the calculated t and F values. Since all significance values for effects and interactions in the last column of Figure 9.10 are greater than 0.05, none of the effects for this example is statistically significant at the 5 percent level of significance. Note that the BC and ABC interactions do not appear in Figure 9.10. They were used to calculate the "no effect" E^2 (see section 8.2).

As another example of statistical software, we will now look at some of the options available using the Jass package. Jass differs from DESIGN-EASE in that it is not menu-driven, so the user must enter commands in response to prompts. However, the columns of the design matrix can be rearranged by the user, so the column assignments used in this book can be exactly dupli-

VARIABLE	PARAMETER ESTIMATE	DF	SUM OF SQUARES	t FOR H0 PARAMETER=0	PROB>|t|
Intercept	57.750000	1		26.411	0.0014
A	5.500000	1	42.000000	2.515	0.1283
B	2.500000	1	50.000000	1.143	0.3713
C	-2.750000	1	60.500000	-1.258	0.3355
AB	0.250000	1	0.500000	0.114	0.9194
AC	-4.000000	1	128.000000	-1.829	0.2089
Std ERROR	2.1866070				

FIGURE 9.10 DESIGN-EASE analysis of variance table

```
Collected by _____

Date _____
```

	Factors			Response
I.D.	cleaner	copper	zinc	y
[1]	1	1	1	51
[2]	1	1	2	71
[3]	1	2	1	51
[4]	1	2	2	69
[5]	2	1	1	49
[6]	2	1	2	50
[7]	2	2	1	58
[8]	2	2	2	63

FIGURE 9.11 Example of Jass report form

cated in Jass. Figure 9.11 is a completed report form for the example in section 3.7. Note that the factor levels are in the standard order used in this book. Jass and other packages will print a report form in either random or standard order. The estimated effects as presented by Jass are given on the left side of Figure 9.12. The estimated effects are plotted in a one-dimensional graph on the right side of that figure. Jass does normal plots and interaction effect plots similar to DESIGN-EASE. Figure 9.13 is a "cube plot" provided by Jass. It is a three-dimensional representation of the responses at each of the eight combinations of levels of the three factors. Each response value in Figure 9.11 appears in Figure 9.13.

2. Replicated full-factorial experiment

All the packages discussed in this chapter can analyze replicated full-factorial experiments with factors each at two levels. To illustrate, the replicated experiment example from section 3.4 was analyzed using the NCSS Statistical System. NCSS interfaces with the user a screen at a time. Figure 9.14 shows an NCSS screen containing thirteen of the sixteen experimental runs from Figure 3.18. The screen does not have room to display all sixteen trials, but the "Count" row near the bottom of the screen indicates that there are a total of sixteen runs in the experiment. Column 5 of Figure 9.15 gives the estimated effects for the example. The estimates for the main effects are consistent with

```
Name     Abbrev    Levels
----     ------    ------
cleaner    c       1    2
copper     C       1    2
zinc       z       1    2


NRuns = 8   NReps = 1   NFactors = 3
```

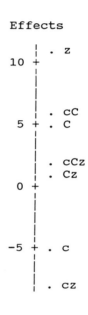

```
                    Effect                      Effects

        Average     57.75
                                            |  . z
                                        10  +
    cleaner (c)     -5.50                   |
    copper  (C)      5.00                   |
      zinc  (z)     11.00                   |
             cC      6.00                   |  . cC
             cz     -8.00                5  +  . C
             Cz       .50                   |
            cCz      1.50                   |
                                            |  . cCz
                                            |  . Cz
                                         0  +
                                            |
                                            |
                                            |
                                       -5   +  . c
                                            |
                                            |
                                            |  . cz
```

FIGURE 9.12 Estimated effects with Jass

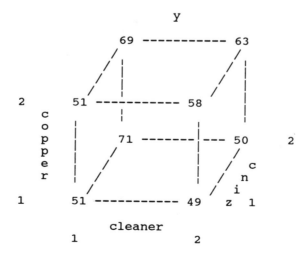

FIGURE 9.13 Cube plot of estimates using Jass

```
===C:\ncss\rep122==================================================Panel 1 ===
Edit value or select a function key.
================================NCSS Editor========================= C S R 1====
                 C1            C2           C3            C4              C5
Row  1           70            60            1           164
Row  2           82            60            1           166             .
Row  3           70            80            1           161             .
Row  4           82            80            1           160             .
Row  5           70            60            2           184             .
Row  6           82            60            2           187             .
Row  7           70            80            2           179             .
Row  8           82            80            2           182             .
Row  9           70            60            1           160             .
Row  10          82            60            1           168             .
Row  11          70            80            1           163             .
Row  12          82            80            1           157             .
Row  13          70            60            2           182             .
Count           (17)          (16)         (16)         (16)            (0)
        ____<Alt>____ ___<Ctrl>___
F1Help        GoTo Var,Row  Special Order║F2Rpt File  Copy Data
F3Open                      Normal Order ║F4Chg RptFl  Paste Data
F5List        List a Row                 ║F6Row Input  Erase Data
F7Labels      Print Labels               ║F8
F9Dir         Enter key                  ║F10Transfer
```

FIGURE 9.14 NCSS screen showing experimental data

those obtained in Figure 3.19. Interaction effects were not calculated in Figure 3.19 because the concept of interactions was not introduced until the next section (section 3.5). When comparing Figures 9.15 and 3.19, keep in mind that Western software generally lists factors in a slightly different order than the ordering used in this book (see Figure 9.3). Figure 9.16 is a normal plot of the estimates and Figure 9.17 is an analysis of variance table. Note that Figure 9.17 indicates that effects number 2, 3 and 4 are all statistically significant at the 0.05 significance level. The normal plot in Figure 9.16 suggests that only effect number 4 appears to be significant. The reason for this discrepancy is that replicated values for the same factor combination tended to be very similar in this experiment. This leads to a very small estimate of variance and hence large F values.

Means / Effects

No.	Effects/ Interactions	Mean −	Mean +	Estimate	Std. Error
	Grand Mean			172.4	0.5
1	A	171.8	173.0	1.3	0.9
2	B	174.4	170.4	−4.0	0.9
3	AB	173.6	171.1	−2.5	0.9
4	C	162.4	182.4	20.0	0.9
5	AC	172.1	172.6	0.5	0.9
6	BC	172.3	172.5	0.3	0.9
7	ABC	171.5	173.3	1.8	0.9

FIGURE 9.15 Estimated effects using NCSS Software System

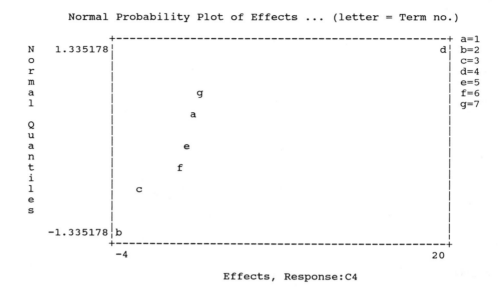

FIGURE 9.16 Normal probability plot of estimated effects using NCSS Software System

Analysis of Variance Report

ANOVA Table for Response Variable: C4

No.	Effects/Interactions	DF	Mean-Square	F-Ratio	Prob>F	Variables
1	A	1	6.25	1.85	0.2107	A: C1
2	B	1	64.00	18.96	0.0024	B: C2
3	AB	1	25.00	7.41	0.0262	C: C3
4	C	1	1600.00	474.07	0.0000	
5	AC	1	1.00	0.30	0.6011	
6	BC	1	0.25	0.07	0.7924	
7	ABC	1	12.25	3.63	0.0932	
	Error	8	3.38			
	Total Sum-Squares	15	1735.75			

FIGURE 9.17 Analysis of variance table from NCSS

3. Eight-run fractional factorial design

Two-level fractional factorial designs are the most useful class of designs for the engineer. All the software packages discussed in this chapter analyze these designs (see Figure 9.2). Section 4.1 gave an example involving an injection molding process. The response variable was a measure of flatness. Four factors were used and eight runs performed. The levels of the factors are given in Figure 4.9. The completed report form for the experiment is in Figure 4.10. Figure 9.18 shows what a user would see on a computer monitor when using the Jass package to set up the injection molding example. Note how, for the most part, the user simply responds to questions. Some other packages also

```
jass> design film

For any query:    HELP    Displays HELP message
                  BACK    Returns to previous prompt
                  QUIT    Exits after default completion of design
                  CMODE   Changes to command mode

Enter number of factors in your design  > 4
Enter maximum number of runs you can do > 8
Enter maximum number of runs per block  > 8

      Selected Design
      ---------------
           8 runs
           4 factors
         1/2 fraction
         No two factor interactions are estimable.

   Effect                Confounded with
   ------                ---------------
   Main effects          3 factor and higher order interactions
   2 factor int.         2 factor and higher order interactions

Do you want to add centerpoints? (default=N) Y/N > n

Do you want to see the alias table for design? (default=Y) Y/N > y
To what level of interaction? (default=2) > 2

   Average

           A
           B
           C
           D
          AB  + CD
          AC  + BD
          AD  + BC

Enter 4 factor names (Default = A  B  C  D)
      Names> CureT MoldT MeltT InjectS

Enter response names (Default = Y)
      Names> Flatness

CureT:        Enter units> sec
              Enter 2 levels> 150 200

MoldT:        Enter units> DegF
              Enter 2 levels> 80 140

MeltT:        Enter units> DegF
              Enter 2 levels> 500 550

InjectS:      Enter units> sec
              Enter 2 levels> 1.00 2.25

Flatness:     Enter units> inches

Enter title for report form.
      Title> Injection Molding Process
```

FIGURE 9.18 Jass Software package; computer monitor display for injection molding example

```
Design name: film
Design title: Injection Molding Process

    Response Abbrev    Units
    -------- ------    -----
  * Flatness    F      inches

    Name   Abbrev        Levels       Units
    ----   ------        ------       -----
    CureT    C        150     200     sec
    MoldT    M         80     140     DegF
    MeltT    T        500     550     DegF
    InjectS  I = CMT  1.00    2.25    sec


    NRuns = 8  NReps = 1  NFactors = 4
    No two factor interactions are estimable.
```

FIGURE 9.19 Jass summary of experimental design

operate in this way. Many experimenters who run only occasional experimental designs find this interactive mode very convenient. Input entered by the user has been underlined. Note in the computer dialogue that the user input the factors in a different order than they appeared in Figures 4.9 through 4.11. Figures 9.19 through 9.21, to be discussed shortly, also list effects in a different order than they were found in Figures 4.9 through 4.11. This reflects the difference in column arrangements between this book and that found in most Western software (see Figure 9.3).

Once the above dialogue with the computer is completed, the user can enter observed response data and generate output similar to Figures 9.11, 9.12, and 9.13 using simple commands. Figures 9.19, 9.20 and 9.21 show some of the output which can be obtained from Jass. Note in Figures 9.18 and 9.19 the caution message regarding confounding of effects. Figure 9.21

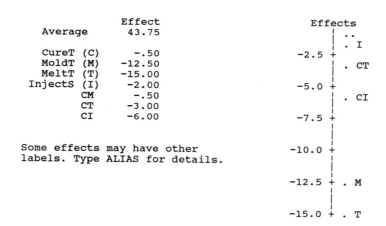

FIGURE 9.20 Jass listing and plot of estimated effect

```
Effect

  43.75        Average

   -.50     CureT  (C)
 -12.50     MoldT  (M)
 -15.00     MeltT  (T)
  -2.00    InjectS  (I)
   -.50            CM  + TI
  -3.00            CT  + MI
  -6.00            CI  + MT
```

FIGURE 9.21 Jass listing of estimated effects with alias structure

shows the confounding relationships. Figure 9.22 is similar output for this
example obtained using NCSS Statistical Software.

4. Sixteen-run fractional factorial design

Use of statistical software can be a real plus when working with sixteen-run
experiments because of the amount of calculation such analyses involve. Consider the example in section 4.3 involving twelve factors and sixteen runs.
The experimental factors and their levels are listed in Figure 4.34. The completed response table is in Figure 4.35. Output from RS/Discover Software is
given in Figures 9.23 and 9.24. Column 3 of Figure 9.23 ("1 Coeff.") contains
the estimated factor effects *divided by 2*. It is necessary to multiply the column 3 values by 2 in order to obtain the actual effect estimates given at the
bottom of Figure 4.35. The one exception to this is the grand mean, which
appears in the first row of the table and does not require multiplication by 2.

```
Two-level Analysis            (Means / Effects)

_____C:\ncss\prob3_____

No.   Effects/Interactions    Mean -     Mean +       Estimate
Std.  Error
      Grand Mean                                        43.8

1     A                        44.0       43.5         -0.5

2     B                        50.0       37.5        -12.5

3     AB+CD                     44.0       43.5         -0.5

4     C                        51.3       36.3        -15.0

5     AC+BD                     45.3       42.3         -3.0

6     BC+AD                     46.8       40.8         -6.0

7     D                        44.8       42.8         -2.0
```

FIGURE 9.22 NCSS listing of estimated effects for a fractional factorial
experiment

```
COEF2   13R x 5C

        Least Squares Coefficients, Response Y, Model DESIGN

    0 Term      1 Coeff.        2 Std. Error   3 T-value   4 Signif.

    -----------------------------------------------------------------
     1  1        165.812500        0.868278
     2  ~A        -0.187500        0.868278       -0.22      0.8429
     3  ~B         3.062500        0.868278        3.53      0.0387
     4  ~C         5.187500        0.868278        5.97      0.0094
     5  ~D         2.937500        0.868278        3.38      0.0430
     6  ~E        -6.437500        0.868278       -7.41      0.0051
     7  ~F         4.062500        0.868278        4.68      0.0184
     8  ~G        -0.312500        0.868278       -0.36      0.7428
     9  ~H         0.687500        0.868278        0.79      0.4863
    10  ~I         4.312500        0.868278        4.97      0.0157
    11  ~J         0.312500        0.868278        0.36      0.7428
    12  ~K         0.687500        0.868278        0.79      0.4863
    13  ~L         7.937500        0.868278        9.14      0.0028


    0 Term      5 Transformed Term

    -----------------------------------------
     1  1
     2  ~A       ((A-1.5)/5e-01)
     3  ~B       ((B-1.5)/5e-01)
     4  ~C       ((C-1.5)/5e-01)
     5  ~D       ((D-1.5)/5e-01)
     6  ~E       ((E-1.5)/5e-01)
     7  ~F       ((F-1.5)/5e-01)
     8  ~G       ((G-1.5)/5e-01)
     9  ~H       ((H-1.5)/5e-01)
    10  ~I       ((I-1.5)/5e-01)
    11  ~J       ((J-1.5)/5e-01)
    12  ~K       ((K-1.5)/5e-01)
    13  ~L       ((L-1.5)/5e-01)

No. cases = 16         R-sq.    =  0.9880       RMS Error = 3.473
Resid. df =  3         R-sq-adj. =  0.9398       Cond. No. = 1
 ~ indicates factors are transformed.
```

FIGURE 9.23 RS/Discover Software output for sixteen-run experiment

Column 5 of Figure 9.23 contains *t* values for the effects. As mentioned earlier, *F* values for the effects are obtained by squaring the *t* values. The level of significance of the *t* and *F* values are given in the "4 Signif." column of Figure 9.23. (Some packages—Jass in particular—provide *t* values but not significance levels for the calculated statistics.) Note that seven of the twelve experimental factors are statistically significant at the 0.05 level of significance (that is, have significance levels less than 0.05). RS/Discover Software provides an exceptionally thorough report on factor confounding, as is given in Figure 9.24. This table was limited by the user to include only main effects and two-factor interactions. The package will also provide higher order interactions upon request. The information in Figure 9.24 is basically the same as

0 Term   1 Desc.   2 Confounding

```
------------------------------------------------------------
 1  A      a        D*L + E*I + F*J + G*K +
 2  B      b        D*K + E*J + F*I + G*L +
 3  C      c        D*J + E*K + F*L + G*I +
 4  D      d        A*L + B*K + C*J + H*I +
 5  E      bcd      A*I + B*J + C*K + H*L +
 6  F      acd      A*J + B*I + C*L + H*K +
 7  G      abd      A*K + B*L + C*I + H*J +
 8  H      abc      D*I + E*L + F*K + G*J +
 9  I      abcd     A*E + B*F + C*G + D*H +
10  J      cd       A*F + B*E + C*D + G*H +
11  K      bd       A*G + B*D + C*E + F*H +
12  L      ad       A*D + B*G + C*F + E*H +
13  A*B    ab       C*H + D*G + E*F + I*J +
14  A*C    ac       B*H + D*F + E*G + I*K +
15  A*D    ad       B*G + C*F + E*H + L + A*
16  A*E    abcd     B*F + C*G + D*H + I + A*
17  A*F    cd       B*E + C*D + G*H + J + A*
18  A*G    bd       B*D + C*E + F*H + K + A*
19  A*H    bc       B*C + D*E + F*G + I*L +
20  A*I    bcd      B*J + C*K + E + H*L + A*
21  A*J    acd      B*I + C*L + F + H*K + A*
22  A*K    abd      B*L + C*I + G + H*J + A*
23  A*L    d        B*K + C*J + D + H*I + A*
24  B*C    bc       A*H + D*E + F*G + I*L +
25  B*D    bd       A*G + C*E + F*H + K + B*
26  B*E    cd       A*F + C*D + G*H + J + B*
27  B*F    abcd     A*E + C*G + D*H + I + B*
28  B*G    ad       A*D + C*F + E*H + L + B*
29  B*H    ac       A*C + D*F + E*G + I*K +
30  B*I    acd      A*J + C*L + F + H*K + B*
31  B*J    bcd      A*I + C*K + E + H*L + B*
32  B*K    d        A*L + C*J + D + H*I + B*
33  B*L    abd      A*K + C*I + G + H*J + B*
34  C*D    cd       A*F + B*E + G*H + J + C*
35  C*E    bd       A*G + B*D + F*H + K + C*
36  C*F    ad       A*D + B*G + E*H + L + C*
37  C*G    abcd     A*E + B*F + D*H + I + C*
38  C*H    ab       A*B + D*G + E*F + I*J +
39  C*I    abd      A*K + B*L + G + H*J + C*
40  C*J    d        A*L + B*K + D + H*I + C*
41  C*K    bcd      A*I + B*J + E + H*L + C*
42  C*L    acd      A*J + B*I + F + H*K + C*
43  D*E    bc       A*H + B*C + F*G + I*L +
44  D*F    ac       A*C + B*H + E*G + I*K +
45  D*G    ab       A*B + C*H + E*F + I*J +
46  D*H    abcd     A*E + B*F + C*G + I + D*
47  D*I    abc      E*L + F*K + G*J + H + D*
48  D*J    c        C + E*K + F*L + G*I + D*
49  D*K    b        B + E*J + F*I + G*L + D*
50  D*L    a        A + E*I + F*J + G*K + D*
51  E*F    ab       A*B + C*H + D*G + I*J +
52  E*G    ac       A*C + B*H + D*F + I*K +
53  E*H    ad       A*D + B*G + C*F + L + E*
54  E*I    a        A + D*L + F*J + G*K + E*
55  E*J    b        B + D*K + F*I + G*L + E*
56  E*K    c        C + D*J + F*L + G*I + E*
57  E*L    abc      D*I + F*K + G*J + H + E*
58  F*G    bc       A*H + B*C + D*E + I*L +
59  F*H    bd       A*G + B*D + C*E + K + F*
60  F*I    b        B + D*K + E*J + G*L + F*
61  F*J    a        A + D*L + E*I + G*K + F*
62  F*K    abc      D*I + E*L + G*J + H + F*
63  F*L    c        C + D*J + E*K + G*I + F*
64  G*H    cd       A*F + B*E + C*D + J + G*
65  G*I    c        C + D*J + E*K + F*L + G*
66  G*J    abc      D*I + E*L + F*K + H + G*
67  G*K    a        A + D*L + E*I + F*J + G*
68  G*L    b        B + D*K + E*J + F*I + G*
69  H*I    d        A*L + B*K + C*J + D + H*
70  H*J    abd      A*K + B*L + C*I + G + H*
71  H*K    acd      A*J + B*I + C*L + F + H*
72  H*L    bcd      A*I + B*J + C*K + E + H*
73  I*J    ab       A*B + C*H + D*G + E*F +
74  I*K    ac       A*C + B*H + D*F + E*G +
75  I*L    bc       A*H + B*C + D*E + F*G +
76  J*K    bc       A*H + B*C + D*E + F*G +
77  J*L    ac       A*C + B*H + D*F + E*G +
78  K*L    ab       A*B + C*H + D*G + E*F +
```

Generators:
E = bcd F = acd G = abd H = abc I = abcd J = cd K = bd L = ad

FIGURE 9.24 RS/Discover Software table of factor confoundings

the information on interactions given in section 4.3 for twelve factors and sixteen runs.

5. Fold-over design

Fold-over designs were the topic of sections 4.2 and 4.4. As you will recall, a fold-over design is a combination of two two-level fractional factorial designs. The design matrix for the second design is obtained by interchanging the 1s and 2s in the design matrix of the first design. The main advantage of fold-over designs is that they allow estimation of main effects free of confounding with second-order interaction effects. The SCA Statistical System software package will be used in this example since it does a particularly nice job of fold-over designs. If you have a package which does not do fold-over designs, you can use two design matrices, as explained at the beginning of this paragraph, and then combine the two sets of estimated effects using the forms in Figure 4.26 and 4.38. Reversing the levels for the second set of trials may require switching the factor levels assigned to "1" and "2" for the first set of trials.

The illustrative example in section 4.2 involves folding over an eight-run design with seven factors. The response table for the first eight runs is given in Figure 4.24. The response table for the second set of eight runs is in Figure 4.28. The estimates obtained in these response tables are combined in Figure 4.29. In order to analyze fold-over experiments using SCA Statistical System, a design matrix containing all sixteen trials is constructed. The trials from the initial eight-run design are listed in the first eight rows of the design matrix, and the trials from the folded design are put in the bottom eight rows of the matrix. The observed response values are added to the matrix as an additional column. If the user requests an analysis based on the first eight rows of the matrix, the estimated effects based on the first eight trials are printed out (see Figure 9.25). Similarly an analysis based on the second set of eight trials (Figure 9.26) or on all sixteen trials—the complete fold-over experiment—can be requested (Figure 9.27). The reader may want to compare Figures 9.25, 9.26, and 9.27 with Figures 4.24, 4.28, and 4.29, respectively. These two sets of figures provide the same factor and interaction estimates, although the order in which they are listed differs because of differences in the order

```
ALIASES OF EACH EFFECT COMPUTED (UP TO ORDER  2 ):

0     99.2875     MEAN

1      -.7250     A + BD + CE + FG
2     -1.8250     B + AD + CF + EG
3      -.7250     C + AE + BF + DG
4     30.1750     D + AB + CG + EF
5     19.8750     E + AC + BG + DF
6     37.3750     F + AG + BC + DE
7     -5.1250     G + AF + BE + CD
--
```

FIGURE 9.25 SCA analysis of first eight trials from a fold-over experiment

```
ALIASES OF EACH EFFECT COMPUTED (UP TO ORDER  2 ):

  0    97.3875      MEAN

  1      -.2750     A - BD - CE - FG
  2    -1.5750      B - AD - CF - EG
  3    -3.7250      C - AE - BF - DG
  4    31.2250      D - AB - CG - EF
  5    18.2750      E - AC - BG - DF
  6   -37.7250      F - AG - BC - DE
  7     2.1750      G - AF - BE - CD
 --
```

FIGURE 9.26 SCA analysis of second eight trials from a fold-over experiment

in which columns of the design matrix are arranged in this book and the order in which factors are handled by the SCA software.

6. Blocked fractional factorial design

Blocking of two-level designs into two or four blocks was discussed in section 4.5. It is also possible to block designs with factors at three or more levels, and to divide experimental trials into three, five, six, or more blocks. Some statistical software does the blocking for you. The report form for recording observed response values may randomize the order of runs within blocks, but keep the blocks separate. If the software does not have a blocking feature, you may be able to identify one factor as the blocking variable, and divide the experiment into blocks based on the values of that one factor. Figure 9.28 contains output from Jass for an eight-run design which has been divided into two blocks. The data in that figure are from an example in section 4.5. The estimated effects in Figure 9.28 are the same as those obtained by hand calculation in Figure 4.41.

```
ALIASES OF EACH EFFECT COMPUTED (UP TO ORDER  2 ):

  0    98.3375      MEAN

  1      -.5000     A
  2    -1.7000      B
  3    -2.2250      C
  4    30.7000      D
  5    19.0750      E
  6      -.1750     F
  7    -1.4750      G
  8      -.5250     AB + CG + EF
  9       .8000     AC + BG + DF
 10      -.1250     AD + CF + EG
 11     1.5000      AE + BF + DG
 12    -3.6500      AF + BE + CD
 13    37.5500      AG + BC + DE
 14      -.2250     BD + CE + FG
 --
```

FIGURE 9.27 SCA analysis of all sixteen runs from a fold-over experiment

```
                    Effect

            41.513        Average

              -.775       Block [1] + cT
              3.775        comp (c) + tg
             -9.175        temp (t) + cg
              9.575        time (T)
              -.825       gsize (g) + ct
              -.525               tT
              -.075               Tg
```

FIGURE 9.28 Jass listing of estimated effects with alias structure for blocked design

7. Use of $\bar{y}$ and *log s* as response variables

Although experimental design software packages will usually allow replication of design points, many packages will not calculate standard deviation (*s*), *log s*, or signal-to-noise ratios. This problem can sometimes be overcome by using software transformation options to calculate these statistics from the raw data values. The statistics can then be analyzed as though they were the original raw response values. As an example, Figure 9.29 contains the same data as Figure 5.13. The data in columns headed "y1," "y2," "y3," "y4," and "y5" contain the raw data. Column "y6" contains average values calculated from the raw data values. Similarly, columns "y7" and "y8" contain values for *s* and *log s*, respectively. The values in the last three columns of Figure 9.29 were obtained using transformations available in Jass. Figure 9.30 contains estimated effects based on the calculated *log s* values. The results are the same as those obtained by hand calculation in Figure 5.15 (allowing for round-off error).

9.4 Summary

The examples in this chapter have demonstrated how easy it is to use computer software when designing and analyzing experiments. Virtually every-

jass> <u>print y1,.,y8</u>

ROW	y1	y2	y3	y4	y5	y6	y7	y8
[1]	73.0	73.2	72.8	72.2	76.2	73.48	1.56589	.1947600
[2]	87.7	86.4	86.9	87.9	86.4	87.06	.70922	-.1492170
[3]	80.5	81.4	82.6	81.3	82.1	81.58	.80436	-.0945485
[4]	79.8	77.8	81.3	79.8	78.2	79.38	1.40784	.1485520
[5]	85.2	85.0	80.4	85.2	83.6	83.88	2.05718	.3132720
[6]	78.0	75.5	83.1	81.2	79.9	79.54	2.92626	.4663130
[7]	78.4	72.8	80.5	78.4	67.9	75.60	5.16769	.7132960
[8]	90.2	87.4	92.9	90.0	91.1	90.32	1.99424	.2997780

FIGURE 9.29 Jass listing of raw data and calculated statistics

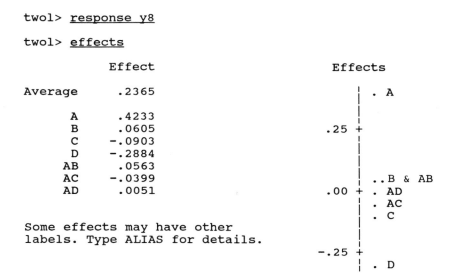

```
twol> response y8

twol> effects

              Effect                      Effects

Average      .2365                   |  . A
                                     |
   A         .4233                   |
   B         .0605                .25 +
   C        -.0903                   |
   D        -.2884                   |
   AB        .0563                   |
   AC       -.0399                   |  ..B & AB
   AD        .0051                .00 + . AD
                                     |  . AC
                                     |  . C
Some effects may have other          |
labels. Type ALIAS for details.      |
                               -.25 +
                                     |  . D
```

FIGURE 9.30 Jass listing of estimated effects using *log s* as the response variable

thing we covered in this book can be done using software packages—everything, that is, except making intelligent decisions regarding confounding of effects, resolution of the design, need for replication or folding over, and analysis of the data. These decisions require an understanding of the material covered in the earlier chapters. But the software takes the "grunt work" out of design and analysis and gives the engineer more time to devote to the interpretation of the experiment.

The software packages presented in this chapter were selected because they represented a cross-section of what is currently on the market. Other good packages are also available. When selecting an experimental design package for your organization, consider compatibility with other data analysis software now being used or soon to be purchased at your company. Also give yourself room to grow—there is more to data analysis than is found in this book, or in any other single book.

10

Using Experiments to Improve Processes

10.1 Engineering Design and Quality Improvement

The Quality profession has come a long way since the days when the Quality Control Department was headed by a "Chief Inspector." But changes in Western perception of quality have not come fast enough. In the 1950s Japanese engineers and managers traveled to the United States to learn how to improve the quality of manufactured products. Today American engineers and managers are visiting Japan to learn the secrets of Japanese quality. But there really are no "secrets." There are just some basic principles which must be followed and tools which must be learned, such as:

- Management cannot be left to the managers. Everyone in the company is a manager of one or more processes, and those closest to a process should participate in its management.
- Random variability is present in all processes. Engineering design and control methods which fail to take randomness in measurements into account lead to out-of-specification products and high production costs.
- Markets cannot be held through use of the latest technology alone. Today customers want proven, reliable, low-cost products designed to meet their needs. Companies must become champions of their customers.
- Experimentation belongs on the manufacturing floor as well as in the research lab. It is naive to assume that the production environment can be exactly reproduced in the lab or that production people cannot be taught methods of experimentation.
- Quality by inspection is no longer a competitive option. Use of Statistical Process Control as the primary method of assuring quality will delay but not prevent a company's demise. The only way to increase market share and profits is by designing quality into all products and processes.

The United States used to loan money to the world. Now it is the largest debtor nation in the world. Today the per capita GNP and wages paid to U.S. workers lag significantly behind those in Japan. In 1982 the United States machine tool business was the biggest in the world. By 1988 the U.S. had sunk to fifth. Many people blame Japan for these dramatic changes. Complaints about the "playing field not being level" are frequently voiced. But on closer scrutiny the problems can be found a little closer to home. The message is clear—many products produced in Western countries cannot compete in world markets because they do not meet customer expectations for quality, cost, and performance. And they will not meet those expectations if Western countries continue to depend on quality control departments for product and process quality. It is time for action:

- Management must reevaluate the way in which they manage. They must provide ongoing leadership in the quality improvement effort. A major obstacle to quality in most companies is a rigid top-down management system which does not encourage process improvement. Dr. Deming's *Out of the Crisis* (1986) is required reading.
- Processes must be standardized, evaluated for capability to meet requirements and brought under statistical control. Statistically designed experiments must be used to improve processes.
- Everyone must be trained in team building, a structured approach to process improvement and tools of data analysis. Employees must become involved in managing and improving processes.

An active quality program focused on designing quality and performance into products and processes is essential for companies which hope to survive into the twenty-first century.

10.2 Steps to Implementing Use of Engineering Design

In an organization where process and product design is already well organized, implementation of the methods presented in this book may involve just incorporating some new techniques into the established procedures. For many companies, however, effective use of this book will require basic changes in their design and development practices. The latter situation will be discussed in this section.

The first implementation project should be done by a small group of three to eight people. If possible, design, development, and manufacturing people should be on the team. All should study this book, preferably as a group, before starting the project. They should also be familiar with the basics of team building. Scholtes' *The Team Handbook* (1988) is a good source. An understanding by the group of the basic concepts of statistical process control would be beneficial as well. It would be good to have one "expert" on the team or as an advisor to the team. By "expert," we mean someone who has taken a one-year college course in statistical experimental design. Use caution here—experimental design has not been used much in quality improvement in many industries, so a very good consultant in quality management or statistical process control may be a neophyte at engineering experimental design.

The following seven steps are based on the seven steps to process improvement recommended in Hitoshi Kume's *Statistical Methods for Quality Improvement*. Experimental design is used primarily in the third step.

Step 1: Project selection

It is probably best to start with an existing process. Pick a production line where quality problems are a fairly common occurrence. Some of the product quality characteristics should be of the variables (continuous) measurement type rather than just go/no go. The production process should have controls which are, or can be, adjusted by operators.

Step 2: Observe the process

Flow chart the process. Identify what characteristics of the process and product are currently being measured or monitored. Collect whatever data are currently available. Determine if the process is in a state of statistical control. If it is not, take steps to remove causes of instability. Compare data on quality characteristics with specifications. If the process is not capable of meeting specifications, reducing process variability should be a top priority for the project team.

Step 3: Analyze the process

Most of the project team's time will be spent in this step. Prepare a cause-and-effect diagram listing possible causes of the process variability you seek to reduce, or listing those noise factors which have the greatest impact on quality. Design one or more experiments to analyze the impact of the critical factors on the chosen quality characteristic(s). Based on the experiment(s), select the key factors to be controlled or adjusted.

Step 4: Take action

Based on the results of the experiments performed in step 3, prepare a plan to implement changes in the process. Present the plan, with supporting data, to the individual or group responsible for the operation of the process. Inform employees working in the process of what changes are to be made, and why. Implement the change. It is usually best to make the change initially on only one line and/or only one shift.

Step 5: Check effectiveness of action

Monitor the process to assess the effectiveness of the change. Whenever possible, involve workers in the evaluation of the process change. Watch for unanticipated side-effects. If the change does not produce the desired results, repeat steps 2 to 5.

Step 6: Standardize the change

Once it has been demonstrated that the process changes have produced the desired results, standardize the changes through training and documentation to make sure the improvements are not lost. This is a critical but often neglected step. Implement the change on all lines or processes. Monitor for conformance.

Step 7. Review and plan

Review the project. What went well? What should be done differently next time? What other opportunities for process improvement using designed experiments were discovered during the project? Are we ready yet to use experimental design in an off-line quality control project for a new product/process, as described in section 2.3, or should another project be done on an existing process?

Appendix A: Summary of Examples

Appendix B: Tables of Random Orderings for Eight- and Sixteen- Run Experiments

The two tables in this appendix can be used to assign random order trial numbers to eight- and sixteen-run fractional factorial experiments. To use these tables, "randomly" select a column from the first (eight runs) or second (sixteen runs) part of this appendix. Enter the numbers in the column selected into the first column of the response table (see Figures 3.25 and 3.54). The experimental trials should then be performed in the order listed in that column of the response table. There are two-hundred sets of random orderings (columns) for eight-run experiments and one-hundred sets for sixteen-run experiments. The tables were generated using the Number Cruncher Statistical System (NCSS) software program developed by Dr. Jerry L. Hintze.

As examples of using this appendix, the orderings of experimental trials shown in Figures 3.11 and 3.27 were obtained from the first and second columns of the table for eight-run experiments, respectively, and the ordering in Figure 3.57 was obtained from the fourth column of the table for sixteen-run experiments.

Table of random orderings for eight-run experiments

Standard order	RANDOM ORDER TRIAL NUMBERS																			
1	4	6	6	5	2	3	7	8	6	4	1	4	4	7	1	7	5	5	5	5
2	1	7	8	8	5	7	6	2	1	3	5	5	7	5	3	6	1	7	8	2
3	8	2	4	3	8	4	3	7	4	8	6	6	5	6	8	4	3	4	4	3
4	5	1	7	6	4	6	8	4	7	7	2	3	6	3	5	5	7	2	1	7
5	6	8	2	2	6	2	4	5	8	1	8	8	8	4	6	1	6	8	3	1
6	3	4	3	4	3	5	1	3	2	2	7	1	3	2	4	2	4	6	7	4
7	2	3	1	1	7	1	5	1	5	5	3	7	1	1	7	8	2	1	2	8
8	7	5	5	7	1	8	2	6	3	6	4	2	2	8	2	3	8	3	6	6
1	6	2	2	5	2	2	8	8	3	3	7	1	8	4	5	2	3	3	2	1
2	3	1	8	3	7	4	5	6	2	1	5	2	3	6	4	3	5	1	1	7
3	2	5	6	4	3	5	1	2	8	2	1	5	5	2	2	4	8	4	7	6
4	7	8	3	6	6	6	7	5	5	4	4	7	6	8	1	5	7	8	5	2
5	4	7	5	1	5	3	2	4	7	6	8	3	2	7	3	8	2	5	8	3
6	8	3	1	2	8	7	3	3	6	8	6	8	7	3	8	7	6	7	6	8
7	5	4	7	8	4	1	4	7	4	5	3	4	1	1	7	1	4	2	4	5
8	1	6	4	7	1	8	6	1	1	7	2	6	4	5	6	6	1	6	3	4
1	8	3	1	4	2	5	4	1	8	1	5	3	4	1	4	1	6	6	6	8
2	5	4	2	3	4	3	5	6	2	3	7	2	1	3	6	3	1	7	4	3
3	6	2	8	6	6	7	2	3	4	2	1	4	6	6	1	5	3	5	2	1
4	7	1	5	8	8	4	8	2	3	5	4	7	2	4	3	8	5	4	3	4
5	4	5	6	1	7	2	3	4	1	4	3	6	5	5	5	4	4	2	1	2
6	1	7	3	2	1	8	7	5	6	8	6	1	3	8	2	2	2	1	8	6
7	2	8	4	5	3	6	6	8	5	7	2	8	7	2	8	6	8	3	5	7
8	3	6	7	7	5	1	1	7	7	6	8	5	8	7	7	7	7	8	7	5
1	6	6	6	5	5	5	6	1	8	4	3	2	3	4	3	4	8	1	5	5
2	8	5	1	2	2	4	3	7	2	6	2	1	2	7	6	6	2	6	6	1
3	3	3	8	4	3	6	2	3	4	2	6	8	1	5	4	7	7	7	7	8
4	1	7	7	3	8	7	1	8	1	3	7	6	6	1	5	8	6	5	4	4
5	5	1	3	1	4	2	4	4	5	8	1	5	4	2	8	1	5	8	3	6
6	2	4	5	6	6	3	5	5	6	7	5	7	8	3	2	2	4	3	2	3
7	4	2	2	8	1	8	8	6	3	1	4	3	5	8	7	5	3	4	1	7
8	7	8	4	7	7	1	7	2	7	5	8	4	7	6	1	3	1	2	8	2
1	6	4	6	6	4	4	6	8	8	4	5	2	6	5	1	2	3	5	2	7
2	7	2	1	2	7	7	5	2	1	7	8	3	7	6	4	4	4	2	1	1
3	4	6	8	4	3	6	7	4	7	3	4	5	5	7	8	3	7	7	5	5
4	3	1	3	3	8	2	4	7	2	2	7	8	2	2	3	5	6	1	7	4
5	8	5	4	8	1	8	8	3	3	6	1	6	4	3	2	6	2	8	8	6
6	1	7	7	5	6	1	1	5	4	1	2	4	1	1	5	8	5	4	3	8
7	5	3	2	1	5	5	2	1	5	8	3	7	8	8	7	1	8	3	4	3
8	2	8	5	7	2	3	3	6	6	5	6	1	3	4	6	7	1	6	6	2

Standard order	RANDOM ORDER TRIAL NUMBERS																			
1	3	4	8	6	1	2	6	8	2	4	4	7	8	6	4	5	1	4	2	4
2	8	8	4	2	5	1	8	6	6	8	8	5	2	8	2	4	3	2	7	2
3	2	6	3	3	2	6	1	7	1	2	2	2	4	3	6	6	2	7	3	1
4	1	5	2	8	8	4	2	4	3	6	7	4	5	2	8	8	7	6	8	5
5	5	7	6	4	6	7	3	3	4	5	5	1	7	5	3	1	8	5	5	7
6	4	3	1	5	3	3	7	1	5	1	6	6	3	4	7	2	5	1	6	3
7	6	1	7	7	7	8	5	2	8	3	3	3	6	1	1	7	6	8	1	8
8	7	2	5	1	4	5	4	5	7	7	1	8	1	7	5	3	4	3	4	6
1	5	3	7	4	4	3	8	8	2	2	4	2	7	8	2	3	1	7	2	1
2	3	2	2	2	1	6	5	3	5	5	6	7	2	6	6	8	5	2	5	6
3	6	5	3	7	5	2	4	7	7	8	3	4	1	5	1	6	6	8	1	7
4	2	4	6	5	2	7	7	4	6	6	5	6	6	1	4	5	8	1	3	4
5	4	1	8	3	3	8	1	6	3	1	8	1	4	7	5	2	7	4	6	3
6	1	6	5	8	6	4	3	1	1	7	2	8	8	4	3	4	2	6	7	5
7	8	8	1	6	8	5	2	2	8	3	1	5	3	2	8	7	4	3	8	2
8	7	7	4	1	7	1	6	5	4	4	7	3	5	3	7	1	3	5	4	8
1	5	7	3	2	3	3	2	4	6	3	1	6	4	4	2	2	2	3	6	2
2	7	5	4	3	4	5	5	7	5	5	6	8	2	2	8	8	7	6	7	6
3	8	1	6	6	7	2	8	6	8	7	8	4	5	3	4	4	5	1	4	5
4	4	3	5	4	5	8	3	2	7	8	3	3	3	6	6	5	8	4	2	3
5	2	4	8	5	2	7	4	1	4	1	2	2	6	1	5	1	6	2	1	8
6	1	2	7	1	8	6	1	8	3	2	5	1	7	7	1	3	1	5	3	1
7	3	6	1	8	6	4	6	3	1	4	7	5	1	8	3	7	3	7	5	4
8	6	8	2	7	1	1	7	5	2	6	4	7	8	5	7	6	4	8	8	7
1	7	2	1	7	4	6	4	1	4	8	6	7	6	4	8	4	2	1	3	3
2	1	7	2	2	3	5	8	4	1	1	5	5	3	6	4	6	6	4	8	6
3	6	6	5	6	5	1	5	7	7	4	3	1	1	8	3	1	7	7	6	4
4	5	8	8	3	8	4	1	2	3	2	4	2	5	1	7	2	8	8	4	5
5	2	5	6	4	7	7	7	6	6	7	2	6	8	2	6	3	5	5	7	2
6	4	1	3	5	1	3	3	3	5	3	8	3	2	5	2	8	1	3	2	1
7	8	4	7	1	2	2	6	8	2	5	1	8	4	7	5	5	4	2	5	8
8	3	3	4	8	6	8	2	5	8	6	7	4	7	3	1	7	3	6	1	7
1	6	5	6	7	2	7	2	3	4	5	6	1	4	1	3	6	4	6	7	5
2	1	4	1	1	5	4	6	6	2	6	3	3	5	8	4	5	8	5	1	8
3	4	7	5	5	3	1	7	7	5	2	8	4	6	5	2	8	5	4	2	6
4	5	8	3	6	6	3	4	4	1	3	2	5	1	7	5	2	3	2	3	1
5	7	6	8	3	7	6	3	2	8	8	1	2	3	2	1	3	2	8	5	2
6	3	1	7	4	1	2	1	5	7	4	7	6	8	4	6	1	7	7	4	3
7	2	3	4	8	8	5	5	1	6	7	5	8	2	3	7	4	6	1	6	4
8	8	2	2	2	4	8	8	8	3	1	4	7	7	6	8	7	1	3	8	7

Table of random orderings for sixteen-run experiments

Standard order	RANDOM ORDER TRIAL NUMBERS														
1	9	2	11	15	14	1	14	16	9	6	1	10	12	9	1
2	15	15	8	4	11	10	6	15	1	15	2	6	16	4	9
3	5	16	14	11	4	9	9	3	10	2	16	3	5	13	3
4	11	8	6	16	5	2	13	12	5	11	5	1	14	1	15
5	13	5	15	2	13	16	1	8	15	3	15	11	6	5	11
6	6	6	4	13	15	13	11	13	16	12	8	12	10	3	7
7	10	4	16	10	9	3	4	14	8	7	9	9	4	16	5
8	4	10	1	9	16	5	3	10	11	14	3	13	1	15	12
9	16	7	13	7	2	15	10	5	7	13	6	16	8	14	16
10	1	11	9	5	3	8	5	2	14	16	12	15	11	10	13
11	7	12	3	1	6	6	15	6	12	8	4	2	15	8	10
12	12	14	10	8	12	4	12	9	2	9	10	14	3	2	2
13	8	1	5	3	8	7	7	11	6	4	14	8	13	6	8
14	3	9	2	12	7	11	16	1	3	1	7	5	7	7	6
15	2	13	7	14	10	14	8	4	4	5	13	4	2	12	14
16	14	3	12	6	1	12	2	7	13	10	11	7	9	11	4
1	13	14	2	11	4	4	10	14	4	9	2	8	2	11	6
2	5	8	15	16	8	3	4	8	10	10	7	4	4	7	13
3	14	12	11	14	16	13	3	5	14	3	4	9	5	3	4
4	3	5	12	5	10	7	5	2	3	4	14	1	16	13	1
5	4	1	3	12	3	11	7	16	15	15	1	14	13	10	12
6	10	7	7	10	15	6	8	9	2	11	13	12	8	9	16
7	1	9	1	4	7	14	14	13	1	12	8	6	10	4	7
8	9	10	14	7	2	10	13	11	13	5	11	10	14	12	10
9	11	2	9	13	12	1	9	12	7	13	15	7	11	15	8
10	6	13	6	9	5	16	11	4	11	16	10	13	6	8	11
11	16	11	8	1	11	2	12	6	12	8	12	15	12	16	2
12	7	15	16	2	14	8	1	10	8	14	9	11	9	2	15
13	8	6	4	3	13	12	6	1	6	2	3	16	15	5	14
14	2	16	10	8	6	5	15	3	9	1	6	3	7	14	5
15	12	3	5	15	1	15	16	7	5	7	5	2	3	6	3
16	15	4	13	6	9	9	2	15	16	6	16	5	1	1	9
1	7	16	15	10	3	10	6	15	3	6	6	15	1	1	5
2	12	15	10	14	15	16	2	6	2	15	15	2	7	5	4
3	16	7	9	5	4	13	16	4	13	2	7	3	2	14	8
4	6	6	12	7	16	11	10	5	8	13	14	10	16	16	7
5	4	4	11	16	1	14	1	13	10	14	11	6	6	15	6
6	11	5	4	15	5	2	3	11	16	4	4	16	12	3	11
7	8	14	5	1	2	8	12	10	7	8	16	14	8	4	14
8	9	11	8	6	9	3	5	9	4	16	1	8	5	7	2
9	5	13	16	3	11	1	11	2	1	7	10	9	11	6	12
10	3	3	14	4	14	12	4	3	11	5	9	7	15	2	9
11	1	1	2	12	6	5	14	16	14	3	13	4	9	10	13
12	13	2	6	13	10	7	9	12	15	12	2	11	3	11	15
13	15	8	7	8	13	15	7	7	12	10	3	13	4	12	16
14	10	12	13	11	7	9	15	8	6	1	8	12	10	8	10
15	14	10	1	2	12	4	8	14	5	9	5	1	13	9	1
16	2	9	3	9	8	6	13	1	9	11	12	5	14	13	3
1	3	13	14	4	11	16	1	13	1	14	5	15	3	10	13
2	4	3	13	5	12	15	8	11	13	13	6	7	11	5	3
3	5	1	15	14	13	4	15	9	9	9	12	2	10	8	7
4	13	5	11	6	1	1	4	14	4	6	3	11	2	14	9
5	10	10	7	10	5	8	13	10	2	4	14	10	8	7	12
6	15	4	4	2	3	3	16	12	11	10	1	16	5	2	4
7	14	8	1	13	8	7	2	15	7	1	7	5	14	4	14
8	12	16	6	9	4	6	12	7	12	12	11	13	4	1	10

Standard order	RANDOM ORDER TRIAL NUMBERS														
9	6	14	3	15	6	14	6	8	16	3	13	6	15	13	16
10	16	9	5	7	7	13	11	4	14	15	9	14	1	15	11
11	1	12	2	3	16	2	14	16	6	2	16	3	12	16	2
12	9	7	9	11	10	9	7	3	8	7	2	12	9	9	6
13	2	11	8	8	9	10	3	2	10	5	15	4	13	3	15
14	11	2	16	1	2	11	9	6	15	16	10	9	6	11	1
15	8	6	10	16	15	5	10	5	3	8	8	8	7	12	8
16	7	15	12	12	14	12	5	1	5	11	4	1	16	6	5
1	3	1	6	11	9	4	11	11	8	9	11	9	12	6	6
2	6	10	11	15	13	15	1	4	2	11	2	8	6	2	7
3	9	13	14	5	1	1	14	6	12	4	7	6	10	8	14
4	5	14	5	14	14	13	4	14	3	5	13	16	1	15	12
5	11	9	13	7	8	6	8	8	11	16	3	11	13	5	8
6	4	16	2	3	7	7	9	12	7	10	6	4	11	9	3
7	1	12	16	4	2	16	2	1	1	8	4	1	2	12	5
8	15	11	15	16	10	12	13	15	15	14	16	12	16	4	1
9	7	3	9	1	5	10	6	10	5	1	12	10	4	11	13
10	13	8	7	2	6	14	12	9	4	12	14	2	7	7	16
11	14	5	10	13	16	8	3	5	13	3	1	13	3	14	15
12	8	7	3	8	15	9	7	16	16	7	15	15	14	3	9
13	10	2	8	9	11	3	5	3	6	15	10	7	15	10	2
14	2	6	4	6	3	5	15	2	14	2	9	5	5	13	10
15	12	15	1	10	4	2	10	7	10	6	8	14	8	1	4
16	16	4	12	12	12	11	16	13	9	13	5	3	9	16	11
1	5	13	7	8	6	4	4	10	9	15	12	8	5	1	9
2	4	12	2	6	3	3	5	12	6	16	3	15	4	8	6
3	9	6	5	13	4	1	2	13	16	2	5	1	13	9	1
4	12	4	14	10	8	8	3	14	12	11	14	6	11	16	16
5	13	11	11	12	5	10	16	11	4	14	4	13	7	7	3
6	15	7	8	9	15	12	10	3	3	13	11	3	12	12	14
7	6	15	9	7	14	2	6	4	15	3	1	16	16	2	13
8	10	5	10	16	11	9	7	2	1	4	13	7	2	14	12
9	3	8	13	4	2	6	1	6	13	1	8	10	6	4	15
10	11	16	4	2	1	11	15	16	11	12	7	14	3	11	10
11	14	1	3	5	9	16	14	8	2	5	2	9	8	10	11
12	8	9	16	1	16	13	12	7	7	6	6	5	1	3	2
13	7	10	6	14	7	5	11	1	8	8	10	12	14	13	5
14	2	14	12	3	10	14	8	5	5	9	15	2	10	5	8
15	1	2	1	15	13	15	9	15	14	7	9	11	15	6	7
16	16	3	15	11	12	7	13	9	10	10	16	4	9	15	4
1	10	4	13	1	1	3	7	10	8	11					
2	3	7	10	8	14	12	15	16	15	4					
3	16	6	9	12	7	11	8	11	4	8					
4	15	16	3	3	6	8	14	13	14	12					
5	12	8	14	4	16	10	4	8	2	9					
6	7	11	2	11	15	4	5	1	5	16					
7	1	3	7	14	8	13	2	3	12	2					
8	11	1	5	16	2	15	9	12	16	7					
9	4	15	11	2	9	6	6	2	6	14					
10	9	10	6	5	4	9	11	9	10	13					
11	5	2	1	13	11	1	3	6	9	10					
12	14	13	4	10	3	5	16	14	1	3					
13	13	5	12	6	12	2	12	4	7	1					
14	8	9	16	9	13	14	1	7	11	5					
15	2	14	15	7	5	16	10	15	13	15					
16	6	12	8	15	10	7	13	5	3	6					

Glossary

alias structure — a listing of the confoundings which occur in an experimental design.

Analysis of Variance (ANOVA) — a statistical methodology for determining the statistical significance of factors included in an experiment.

blocking — grouping trials of an experiment into subgroups or "blocks." Trials in the same block are performed at the same time, or have some other common characteristic with respect to their being performed.

confounding — if an experiment is designed in such a way that two effects (main, interaction or block) cannot be estimated independently of each other, then the two effects are said to be confounded.

control factors — factors used in an experiment which can be controlled during the experiment and also during normal production.

customer relations — the second stage of Taguchi's on-line quality control system. During this stage services are provided to customers and information on product field problems is used to improve product and manufacturing process designs.

degrees of freedom — a name frequently given to a parameter of a probability distribution. Degrees of freedom is often a function of sample size.

design parameters — see control factors.

effect — a factor effect is the change in average response when the factor goes from its low level to its high level.

external noise — variation in environmental factors which cannot be controlled and which may affect product characteristics.

F distribution — a probability distribution used in Analysis of Variance.

fold-over design — a two-level experimental design obtained from another two-level design by interchanging all the 1s and 2s (or −1s and 1s) in the original design matrix. Folding over a design always results in a design of higher resolution.

factorial design — see full factorial design.

fractional factorial design — an experimental design in which some of the effects are confounded with each other.

full factorial design — an experimental design in which each possible combination of factors is included once. Full factorial designs have no confounded effects.

inner array — an experimental design for factors which are all controllable.

interaction effect — a measure of the extent to which two or more factors jointly influence the way they each affect a response variable.

internal noise — variation in a product or production process characteristic caused by wear or other factors related to age or length of use.

larger-is-better quality characteristic — a product or process characteristic which the experimenter wants to maximize. Yield, bursting strength of packaging and bond strength are examples.

log s — the logarithm (usually base 10 or base e) of a sample standard deviation.

loss function — the cost or loss associated with a given quality characteristic value expressed as a function of distance from a target value. The loss function may involve specification limits.

main effect — see effect.

"mu" (μ) — a Greek letter usually used to denote the average value of a product or process quality characteristic based on many such products or over an extended period of time. This is the average for all the measurements of a population.

noise factor — an uncontrollable source of variation in the functional characteristics of a product or production process.

nominal-is-best quality characteristic — a characteristic which has a specific target value which optimizes product performance. A measurement such as the diameter of a shaft would have a target value.

normal distribution — a probability distribution for populations of random measurements having a "bell-shaped" pattern to their distribution.

normal plot — a graphical technique for plotting points on special graph paper to aid in identifying unusual values of measurements or estimated effects.

off-line quality control — that part of a company's quality system which is concerned with designing quality into products and production processes.

on-line quality control — that part of a company's quality system which is concerned with meeting or exceeding specification developed during off-line quality control activities and responding to customer dissatisfaction.

orthogonality — two effects are orthogonal if neither estimated effect is affected or biased by the other. If two columns of numbers have the property that the sum of the products of their respective terms is equal to zero, the columns are orthogonal and the estimated effects based on those columns are orthogonal.

outer array — an experimental design for factors which are all noise factors.

parameter design — the second step of each of Taguchi's two stages of off-line quality control. During the parameter design step the optimal settings for product or production process parameters are determined.

process design — the second stage of Taguchi's two-stage system of off-line quality control. During process design, production and process engineers develop manufacturing processes to meet specifications developed during the product design stage.

product design — the first stage of Taguchi's two-stage system of off-line quality control. During product design a new product is developed or an existing product is modified.

production quality control methods — the first stage of Taguchi's on-line quality control system. It has three forms: process diagnosis and adjustment, prediction and correction, and measurement and action.

quality of conformance — manufacturing products or providing services which meet previously determined specifications. Quality of conformance is basically the same as Stage 1 of Taguchi's on-line quality control.

quality of design — activities which assure that new or modified products or services are designed to meet customer needs and expectations, and are economically achievable. Quality of design is basically the same as Taguchi's off-line quality control.

random experiment — an experiment for which the observed outcomes cannot be exactly predicted in advance. Almost all engineering experiments are random.

randomization — performing experimental trials in a random order, rather than in the order in which they are logically listed in, say, a response table. Randomization helps protect against unknown factors which might influence response values and bias estimated effects.

replicated experiment — an experiment in which a given experimental design is carried out more than once.

report form — a form which contains a listing of the experimental trials to be performed, including the levels for the experimental factors, plus space to record the observed responses. The experimental trials should be listed in the order in which they are to be performed.

resolution III designs — experimental designs in which main effects are not confounded with main effects, but one or more main effects are confounded with two-factor interaction effects.

resolution IV designs — experimental designs in which main effects are not confounded with main effects or with two-factor interactions, but two or more two-factor interaction effects are confounded with each other.

resolution V designs — experimental designs in which no main effects and two-factor interactions are confounded with each other, but at least one two-factor interaction effect is confounded with a three-factor interaction effect.

response table — a form used to simplify the calculation of effects. The observed response values are written in a column of the response table, and calculations performed as suggested by the table layout.

response variable — a measurable product or process quality characteristic which is affected by other factors. In an experiment, various levels are set for the experimental factors and the resulting value for the response variable is then measured.

robust — a product or manufacturing process design is robust if it is relatively insensitive to noise factors which are present.

"**sigma**" (*σ*) — a Greek letter usually used to denote the standard deviation of a product or process quality characteristic for many such products for an extended period of time. This is the standard deviation for all the measurements of a population.

signal-to-noise ratio — a class of statistics used as measures of the effect of noise factors on performance characteristics.

small-is-better quality characteristic — a product or process characteristic which the experimenter wants to minimize. Distortion, scrap and delivery time are examples.

standard deviation — a statistical measure of variability in a sample or population.

statistic — a function of sample data.

system design — the first step of each of Taguchi's two stages of off-line quality control. During the system design step prototype product designs and manufacturing designs are determined based on proven technology. Use of existing equipment and low cost parts are emphasized at this step.

target value is best — see nominal is best.

tolerance design — the third step of each of Taguchi's two stages of off-line quality control. During the tolerance design step tolerances are established for product or manufacturing process characteristics.

unit-to-unit noise — differences in products built to the same specifications caused by variability in materials, manufacturing equipment and assembly processes.

variability — the measurable effects of noise on products and processes.

Bibliography

American Society for Quality Control (1983). *Glossary and tables for statistical quality control*. Milwaukee, WI: ASQC.

Barker, T. B. (1985). *Quality by experimental design*. Dekker.

Barker, T. B. (1986). "Quality engineering by design; Taguchi's philosophy." *Quality Progress, 18*, December, 32–42.

Box, G. E. P. (1988). "Signal-to-noise ratios, performance criteria and transformations." *Technometrics, 30*, 1–40.

Box, G. E. P., Bisgaard, S., and Fung, C. (1988). "An explanation and critique of Taguchi's contributions to quality engineering." *Quality and Reliability Engineering International,* (May).

Box, G. E. P. and Draper, N. R. (1987). *Empirical model-building and response surfaces*. New York: John Wiley.

Box, G. E. P. and Hunter, J. S. (1961). "The 2^{k-p} fractional factorial designs." *Technometrics, 3*, 311–351 (Part I), 449–458 (Part II).

Box, G. E. P., Hunter, W. G., and Hunter, J. S. (1978). *Statistics for experimenters*. New York: John Wiley.

Caplan, F. (1980). *The quality system: A sourcebook for managers and engineers*. Radnor, PA: Chilton.

Carrothers, H. S. (1985). "Non linear spark control valve sonic welder experiment." *Third Supplier Symposium on Taguchi Methods*, pp. 410–436. Dearborn, MI: American Supplier Institute.

Crosby, P. B. (1979). *Quality is free*. New York: McGraw-Hill.

Daniel, C. (1959). "Use of half-normal plot in interpreting factorial two-level experiments." *Technometrics, 1*, 311–341.

Daniel, C. (1976). *Applications of statistics to industrial experiments*. New York: Hafner (Macmillan).

Deming, W. E. (1986). *Out of the crisis*. Cambridge, MA: Massachusetts Institute of Technology, Center for Advanced Engineering Study.

Desrochers, G. and Ewing, D. (1984). "Leaf spring free height analysis using Taguchi methods." *Second Supplier Symposium on Taguchi Methods*, pp. 38–47. Dearborn, MI: American Supplier Institute.

Draper, N. R. and Smith, H. (1981). *Applied regression analysis*, 2nd ed. New York: John Wiley.

Feigenbaum, A. V. (1983). *Total quality control*, 3rd ed., New York: McGraw-Hill.

Filley, R. D. (1985). "2,500 IEs work for worldwide Philips, a leader in quality and technology." *Industrial Engineering*, October, 96–104.

Guttman, I., Wilks, S. S., and Hunter, J. S. (1971). *Introductory engineering statistics*, 2nd ed., New York: John Wiley.

Hauser, J. R. and Clausing, D. (1988). "The house of quality." *Harvard Business Review*, May–June, 63–73.

Imai, M. (1986). *Kaizen: The key to Japan's competitive success*. New York: Random House.

Ireson, W. G. and Coombs, C. F., Jr., editors (1988). *Handbook of reliability engineering and management*. New York: McGraw-Hill.

Jones, B. A. (1988). "A robust approach to Taguchi Methods." *12th Annual Rocky Mountain Quality Conference*, June 8, 1988.

Juran, J. J. (1964). *Managerial breakthrough*, New York: McGraw-Hill.

Kackar, R. N. (1985). "Off-line quality control, parameter design and the Taguchi method." *Journal of Quality Technology*, *17*, 176–188.

Kackar, R. N. (1986). "Taguchi's quality philosophy: Analysis and commentary." *Quality Progress*, *19* (December), 21–29.

Kackar, R. N. and Shoemaker, A. C. (1986). "Robust design: A cost-effective method for improving manufacturing processes." *AT&T Technical Journal*, *65*, 2 (March–April), 39–50.

Kume, H. (1987). *Statistical methods for quality improvement*.

Lochner, R. H. and Matar, J. E. (1988). "Enhanced Taguchi methods using Western designs." *ASQC Quality Congress Transactions*, pp. 666–671.

Miller, R. (1988). "Continuing the Taguchi tradition." *Managing Automation* (February), 34–36.

Nachtsheim, C. J. (1987). "Tools for computer-aided design of experiments." *Journal of Quality Technology*, *19*, 132–160.

Ott, E. R. (1975). *Process quality control: Troubleshooting and interpretation of data*. New York: McGraw-Hill.

Ott, L. (1988). *An introduction to statistical methods and data analysis*, 3rd ed. Boston: P.W.S.-Kent.

Phadke, M. S. (1986). "Design optimization case studies." *AT&T Technical Journal*, *65*, 2 (March–April), 51–68.

Phadke, M. S. (1986). *Quality engineering using robust design*, Englewood Cliffs, NJ: Prentice-Hall.

Pignatiello, J. J., Jr. and Ramberg, J. S. (1985). "Discussion on off-line quality control, parameter design, and the Taguchi method by R. N. Kackar." *Journal of Quality Technology*, *17*, 198–206.

Plackett, R. L. and Burman, J. P. (1946). "The design of optimum multifactorial experiments." *Biometrika*, *33*, 305–325.

Quality Progress (1988). Five articles on quality function deployment *21* (June).

Quality Progress (1989). "1989 QA/QC software directory." 22 (March), 24–64.

Quinlan, J. (1985). "Product improvement by application of Taguchi methods." *Third Supplier Symposium on Taguchi Methods*, pp. 367–384. Dearborn, MI: American Supplier Institute.

Ross, P. J. (1988). *Taguchi techniques for quality engineering.* New York: McGraw-Hill.

Scholtes, P. R. (1988). *The team handbook: How to use teams to improve quality.* Madison, WI: Joiner Associates.

Sullivan, L. P. (1986). "Quality function deployment." *Quality progress, 19* (June), 39–50.

Taguchi, G. (1986). *Introduction to quality engineering.* Tokyo: Asian Productivity Organization.

Taguchi, G. (1987). *System of experimental design, vols. 1 and 2.* White Plains, NY: Quality Resources; and Dearborn, MI: American Supplier Institute.

Taguchi, G., Elsayed, E., and Hsiang, T. C. (1988). *Quality engineering in production systems.* New York: McGraw-Hill.

Walton, M. (1986). *The Deming management method.* Milwaukee, WI: Quality Press.

Winner, R. I., Pennell, J. P., Bertrand, H. E., and Slusarczuk, M. M. G. (1988). "The role of concurrent engineering in weapons systems acquisition." IDA Report R-338, Alexandria, VA: Institute for Defense Analysis.

Index

Acknowledgment of Permissions

The following have generously granted permission to use copyright material.

The American Supplier Institute, P.O. Box 567, Dearborn, Michigan, 48121-9567; sponsor of the ASI Supplier Symposiums on Taguchi Methods.

BBN Software Products Corporation, 10 Fawcett Street, Cambridge, Massachusetts, 02138; publisher of RS/Discover Software.

Dr. Jerry L. Hintze, 1000 East 329 North, Kaysville, Utah, 84037; publisher of the Number Cruncher Statistical System (NCSS).

Joiner Associates, Inc., 3800 Regent Street, P.O. Box 5445, Madison, Wisconsin, 53705-0445; publisher of Jass software.

Bradley Jones, author of "A Robust Approach to Taguchi Methods," presented at the 12th Annual Rocky Mountain Quality Conference, June, 1988.

R. N. Kackar, author of "Taguchi's Quality Philosophy," which appeared in *Quality Progress*.

Quality Progress/The American Society for Quality Control, 310 West Wisconsin Avenue, Milwaukee, Wisconsin, 53203.

Scientific Computing Association, Lincoln Center, Suite 106, 4513 Lincoln Avenue, Lisle, Illinois, 60523; publisher of the SCA Statistical System: Quality Improvement Package.

STAT-EASE, Inc., 3801 Nicollet Avenue South, Minneapolis, Minnesota, 55409; publisher of DESIGN-EASE software.